Carl Ernst Bock

Atlas of Human Anatomy

Carl Ernst Bock

Atlas of Human Anatomy

ISBN/EAN: 9783742862884

Manufactured in Europe, USA, Canada, Australia, Japa

Cover: Foto ©berggeist007 / pixelio.de

Manufactured and distributed by brebook publishing software
(www.brebook.com)

Carl Ernst Bock

Atlas of Human Anatomy

HUMAN ANATOMY,

WITH EXPLANATORY TEXT.

BY

PROF. DR. C. E. BOCK (Leipsic).

CONTAINING

Thirty-eight Coloured Plates

OF

THE BONES, MUSCLES, VESSELS AND NERVES OF THE HUMAN BODY,

ORGANS OF SENSE, EYE, EAR, NOSE, AND TONGUE,

RESPIRATORY APPARATUS, ABDOMINAL AND PELVIC VISCERA, ORGANIZATION OF FŒTUS, THE TEETH,

WITH THE GENITO-URINARY ORGANS OF THE MALE AND FEMALE.

———————

NEW YORK:

WILLIAM WOOD & COMPANY, 27 GREAT JONES STREET.

1879.

PREFACE.

THIS Atlas of Prof. Dr. C. E. BOCK has been most favorably received in Europe; its success being such as to call for a Sixth Edition. It has been translated into several languages, and wherever used has met with approval. It was published for the first time in London last year, and this Edition is reprinted from the English translation.

An Atlas of Human Anatomy, of moderate dimensions and corresponding cost, has long been a desideratum in the profession; not only with those who have been engaged in the study of Medicine, but also with such as, in active practice, were called on to perform serious operations at a short notice and desired to refresh their knowledge by appropriate means. To both, this work will be found a valuable acquisition, not only from its excellent delineation of the various parts of the human body, but from the exceeding accuracy of the details represented.

27 GREAT JONES STREET, NEW YORK.

CONTENTS.

8

CONTENTS.

10CONTENTS.

PLATE I.

BONES OF HEAD.

Fig. 1.

Front view of Cranium.

A. Os frontis.
B. Os parietale.
C. Great wing of sphenoid bone.
D. Os temporis.
E. Malar bone.
F. Superior maxillary bone.
G. Nasal bone.
H. Inferior maxillary bone.
a. Coronal suture.
b. Frontal suture.
c. Squamous suture.
d. Frontal crest.
e. Superciliary arch.
f. Glabella.
g. Zygomatic process of malar bone.
h. Supra-orbital ridge.
i. Supra-orbital foramen.
k. Nasal process.
l. Frontal crest of temporal ridge.
m. Nasal or ascending process of superior maxilla.
n. Malar process of superior maxilla.
o. Alveolar process of superior maxilla.
p. Infra-orbital foramen.
q. Superior maxillary fossa.
r. Nasal spine of superior maxilla.
s. Anterior nasal opening.
t. Infra-orbital margin of superior maxilla.
u. Fossa of lachrymal sac.
v. Alveolar process of superior maxilla.
w. Maxillary process of malar bone.
x. Frontal process of malar bone.
y. Temporal process of malar bone.
z. Anterior malar foramen.
α. Mental protuberance of inferior maxilla.
β. Anterior opening of inferior maxillary canal.
γ. Angle of inferior maxilla.
δ. Ascending ramus of inferior maxilla.
ε. Mastoid process.
ζ. Foramen opticum.
η. Superior orbital fissure of foramen lacerum anterius.
θ. Inferior orbital fissure of foramen lacerum anterius.
ι. Posterior malar foramen.

Fig. 2.

Lateral or side view of Cranium.

A. Os frontis.
B. Os parietale.
C. Os temporis.
D. Great wing of sphenoid.
E. Malar bone.
F. Superior maxilla.
G. Nasal bone.
H. Inferior maxilla.
a. Frontal eminence.
b. Superciliary arch.
c. Glabella.
d. Nasal process of frontal bone.
e. Supra-orbital margin of frontal bone.
f. Supra-orbital foramen of frontal bone.
g. Malar process of frontal bone.
h. External frontal crest of temporal ridge.
i. Temporal or semicircular ridge.
k. Coronal suture.
l. Parietal eminence.

m. Squamous plate of os temporis.
n. Mastoid prolongation.
o. Mastoid process.
p. Mastoid fossa.
q. Meatus auditorius externus.
r. Zygomatic arch.
s. Temporal or horizontal process of malar bone.
t. Frontal or ascending process of malar bone.
u. Maxillary process of malar bone.
v. Anterior malar foramen.
w. Malar process of superior maxilla.
x. Superior maxillary fossa.
y. Infra-orbital foramen.
z. Superior maxillary protuberance.
1. Nasal process of superior maxilla.
2. Alveolar process of superior maxilla.
3. External pterygoid process of sphenoid bone.
4. Mental protuberance of inferior maxilla.
5. Inferior mental spine.
6. Anterior opening of inferior maxillary canal.
7. External oblique line of inferior maxilla.
8. Alveolar border of inferior maxilla.
9. Angle of inferior maxilla.
10. Coronoid process of inferior maxilla.
11. Condyloid process of inferior maxilla.
12. Semilunar notch of inferior maxilla.
13. Anterior nasal opening.
14. Anterior nasal spine.
15. Fossa for lachrymal sac.
16. Squamous suture.
17. Mastoid suture.
18. Transverse suture.

Fig. 3.

Vertical section of Facial Bones, showing inner surface of orbit, antrum Highmorianum, and lateral surface of superior maxilla, with portions of sphenoid, temporal, and palate bones posteriorly (spheno-maxillary fossa).

A. Os frontis.
B. Os nasale.
C. Superior maxilla.
D. Palate bone (perpendicular portion).
E. Ethmoid bone.
F. Lachrymal bone.
G. Sphenoid bone.
a. Fossa for lachrymal sac.
b. Infra-orbital foramen.
c. Infra-orbital canal.
d. Ethmoidal foramina.
e. Antrum Highmorianum or superior maxillary sinus.
f. Pterygoid process.
g. Pterygo-palatine canal.
h. Spheno-palatine foramen.
i. Orbital process of palate bone.
k. Sphenoidal process of palate bone.
l. Orbital process of frontal bone.
m. Anterior clinoid process.
n. Sella turcica.
o. Foramen opticum.
p. Posterior clinoid process.
q. Carotid canal for interior carotid artery.
r. Lingual foramen.
s. Vidian canal.
t. Styloid process of temporal bone.

Fig. 4.

Vertical section of Facial Bones, showing interior and outer wall of nasal cavity, with portions of frontal, ethmoidal, and sphenoidal sinuses.

A. Os frontis.
B. Body of sphenoid bone.
C. Pterygoid process.
D. Vertical plate of palate bone.
E. Horizontal or palate process of palate bone.
G. Inferior spongy bone.
H. Nasal plate of ethmoidal spongy bones.
I. Ascending plate of superior maxilla.
K. Nasal bone.
a. Frontal sinus.
b. Ethmoidal sinus.
c. Sphenoidal sinus.
d. Superior spongy bone.
e. Middle spongy bone.
f. Spheno-palatine foramen.
g. Internal pterygoid process.
h. External pterygoid process.
i. Opening of nasal duct or lachrymal canal.
k. Incisive canal.
l. Groove for ethmoidal nerve.

Fig. 5.

Os Ethmoides, cerebral or upper surface.

A. Cribriform plate.
B. Ethmoidal cells.
C. Vertical or nasal plate.
a. Crista galli.
b. Hamular process of crista galli.
c. Foramina cribrosa.
d. Orbital plate.

Fig. 6.

Inferior or nasal surface of Ethmoid Bone.

A. Superior nasal meatus.
B. Vertical or nasal plate.
C. Cribriform plate.
a. Hamular process of ethmoid bone.
b. Middle spongy bone.
c. Superior spongy bone.

Fig. 7.

Os Palati or Palate Bone, inner or nasal surface.

A. Perpendicular or nasal plate.
B. Horizontal plate.
a. Nasal crest.
b. Interior nasal spine.
c. Inferior turbinal crest.
d. Superior turbinal crest.
e. Orbital process.
f. Sphenoidal process.
g. Spheno-palatine foramen.
h. Nasal process.
i. Pyramidal process.

Fig. 8.

Os Hyoides, anterior aspect.

a. Body.
b. Cornua majora.
c. Cornua minora.
d. Knob of greater cornu for thyro-hyoid ligament.

Tab. I.

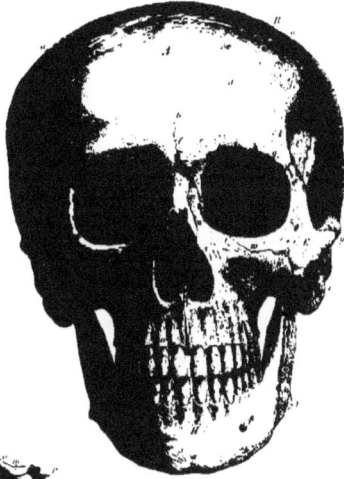

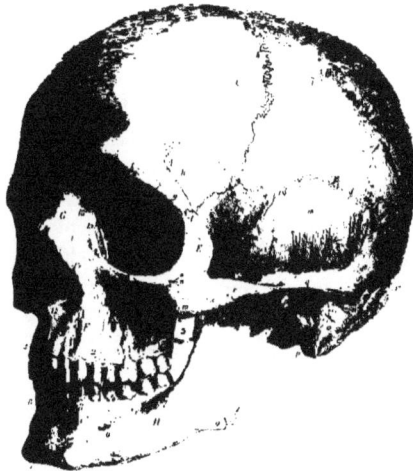

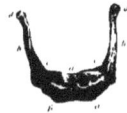

PLATE II.

BONES OF HEAD (continued).

Fig. 1.

Base of Skull—inner or cerebral surface.

A. Os frontis (section of).
B. Lesser wings of sphenoid bone.
C. Greater wings of sphenoid bone.
D. Squamous plate of temporal bone.
E. Petrous portion of temporal bone.
F. Mastoid portion of temporal bone.
G. Occipital bone (section of).
H. Basilar or cuneiform process of occipital bone.
1. Ethmoid bone (cribriform plate).
a. Orbital plates of frontal bone.
b. Digital depression.
c. Internal frontal spine.
d. Crista galli.
e. Foramina cribrosa.
f. Anterior clinoid processes.
g. Optic foramen.
h. Middle clinoid processes.
i. Sella turcica.
k. Posterior clinoid processes.
l. Internal carotid sulcus.
m. Foramen lacerum anterius orbitale.
n. Foramen rotundum.
o. Foramen ovale.
p. Foramen spinosum.
η. Hiatus canalis Fallopii.
r. Meatus auditorius internus.
s. Aqueduct of vestibule.
t. Jugular foramen.
u. Fossa occipitalis.
v. Processus anonymus.
w. Ante-condyloid foramen.
x. Posterior condyloid foramen.
y. Mastoid foramen.
z. Foramen magnum.
а. Sigmoid fossa.
β. Internal occipital protuberance.
γ. Internal occipital crest.
1. Foramen cæcum.
ε. Clivus Blumenbachii.
ζ. Fossa cerebelli.
η. Lingula.
3. Sulcus transversus.
ι. Condyloid occipital eminence.

Fig. 2.

Inferior surface of Cranium—base of skull.

A. Bony or hard palate.
B. Alveolar ridge of superior maxilla.
C. Superior maxilla.
D. Palate or horizontal plate of palate bone.
E. Pterygoid processes of sphenoid bone.
F. Greater wing of sphenoid.
G. Vomer.
H. Squamous plate of temporal bone.
I. Mastoid process.
K. Petrous portion of temporal bone.
L. Basilar process.
M. Condyloid part of occipital bone.
O. Zygomatic arch.
a. Foramen incisivum.
b. Posterior nasal spine.
c. Foramina palatina.
d. Internal pterygoid plate and hamular process.
e. Pterygoid fossa.
f. External pterygoid plate.
g. Posterior nasal openings.
h. Foramen ovale.

i. Foramen spinosum.
k. Inferior orbital or spheno-maxillary fissure.
l. Articular fossa of temporal bone.
m. Condyloid eminence or transverse root of zygoma.
n. Fissura Glaseri.
o. Eustachian tube.
p. Meatus auditorius externus.
q. Internal carotid foramen.
r. Styloid process.
s. Style-mastoid foramen.
t. Aqueduct of cochlea.
u. Jugular foramen or foramen lacerum in basi cranii.
v. Fossula petrosa.
w. Condyles of occipital bone.
x. Anterior condyloid foramen.
y. Posterior condyloid foramen.
z. Mastoid process.
α. Mastoid fossa.
β. Mastoid foramen.
γ. Posterior occipital ridge.
δ. Posterior occipital eminence.
ε. Inferior semicircular line.
ζ. Superior semicircular line.
η. Spinous process.

Fig. 3.

Os Temporis, Temporal bone—external surface. (*Vide* Plate I., Fig. 2.)

A. Squamous plate.
B. Mastoid portion.
C. Petrous portion.
a. Zygomatic process.
b. Condyloid process or transverse root of zygoma.
c. Glenoid cavity or articular fossa.
d. Fissura Glaseri.
e. Meatus auditorius externus.
f. Mastoid process.
g. Mastoid foramen.
h. Mastoid fossa.
i. Styloid process.

Fig. 4.

Os Temporis—inner or cerebral surface. (*Vide* Fig. 1.)

A.—C. as Fig 3.
a. Digital depressions.
b. Juga cerebralia.
c. Zygomatic process.
d. Mastoid process.
e. Sigmoid fossa.
f. Mastoid foramen.
g. Styloid process.
h. Opening of carotid canal.
i. Meatus auditorius internus.
k. Petrosal ridge.
l. Aqueduct of vestibule.
m. Eminentia arcuata.

Fig. 5.

Os Sphenoides, Sphenoid Bone—inner or cerebral surface. (*Vide* Fig. 1.)

A. Body.
B. Lesser wings.
C. Greater wings.
a. Middle clinoid processes.
b. Sella turcica.
c. Posterior clinoid processes.

d. Sulcus caroticus.
e. Anterior clinoid processes.
f. Foramen opticum.
g. Superior orbital fissure or foramen lacerum anterius.
h. Foramen rotundum.
i. Foramen ovale.
k. Foramen spinosum.
l. Processus spinosus.
m. Lingula.
n. Ethmoidal process.
o. Superior articulating surface.
p. Posterior articulating surface.
q. Posterior inferior surface.
r. Clivus.

Fig. 6.

Os sphenoides—anterior surface.

A. Body.
B. Lesser wings.
C. Greater wings (orbital surface).
D. Pterygoid processes.
a. Posterior clinoid processes.
b. Sella turcica.
c. Sphenoidal sinuses.
d. Nasal spine.
e. Superior orbital fissure.
f. Orbital surface of greater wing.
g. Foramen rotundum.
h. Vidian canal.
i. Spinous process.
k. Pterygoid groove for pterygo-palatine canal.
l. External pterygoid process.
m. Internal pterygoid process.
n. Pterygoid cleft.
o. Hamular process of internal pterygoid plate.

Fig. 7.

Maxilla Inferior—outer or anterior surface.

A. Body.
D. Ascending ramus.
a. Base or inferior margin.
b. Alveolar border.
c. Mental protuberance.
d. Mental foramen.
e. External oblique line.
f. Angle.
g. Condyloid process.
h. Coronoid process.
i. Sigmoid notch.
k. Incisor teeth.
l. Cuspidati.
m. Molar teeth.

Fig. 8.

Maxilla Inferior—posterior or inner surface.

A. Body.
B. Ascending ramus.
a. Base or inferior margin.
b. Alveolar border.
c. Internal mental spine.
d. Internal oblique line.
e. Angle.
f. Condyloid process.
g. Coronoid process.
h. Sigmoid notch.
i. Inferior maxillary canal (internal opening)
k. Mylo-hyoid groove.

Tab. II.

Fig.4.

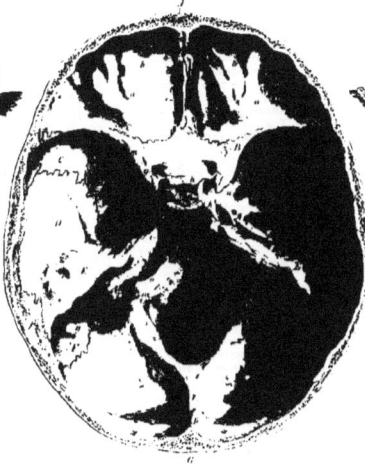

Fig.1.

Fig.5.

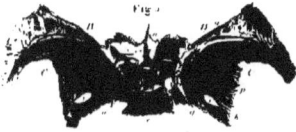

Fig.3.

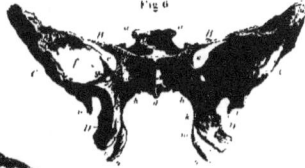

Fig.6.

Fig.2.

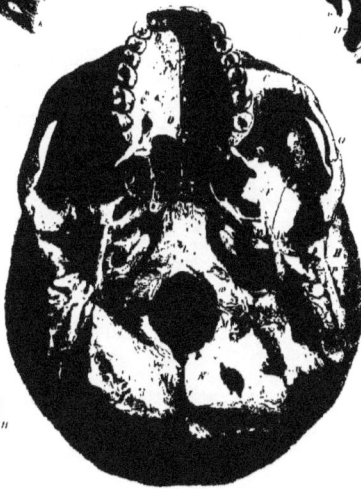

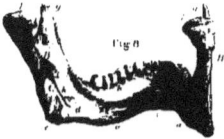

Fig.8.

Fig.7.

PLATE III.

BONES OF TRUNK.

Fig. 1.

Spine (vertebræ), Thorax, Clavicle, and portion of Scapula.

a   Atlas or first vertebra.
b   Dentata or second vertebra.
c   Last cervical vertebra.
d   Vertebral canal for vertebral artery.
e   Odontoid process.
f   First dorsal vertebra.
g   Last dorsal vertebra.
h   First lumbar vertebra.
i   Last lumbar vertebra.
k   First rib.
l   Last true or sternal rib.
m   First false or asternal rib.
n   Last floating rib.
o   Manubrium or first bone of sternum.
p   Body or middle piece of sternum.
q   Ensiform or xiphoid process.
r   Clavicle.
s   Scapula.
t   Glenoid cavity of scapula.

Fig. 2.

Pelvis.

A.   Os sacrum.
B.   Os innominatum.
C.   Os ilium.
D.   Os ischii.
E.   Os pubis.
a   Superior oblique or articular process of sacrum.
b   Base or promontory of sacrum.
c   Linear arcuata interna.
d   Anterior sacral foramina.
e   Internal semicircular ridge or inferior brim of pelvis.
f   Sacro-iliac symphysis.
g   Crest of ilium or superior brim of pelvis.
h   Anterior superior spinous process of ilium.
i   Anterior inferior spinous process of ilium.
k   Anterior semilunar notch.
l   Spine of ischium.
m   Ilio-pubal eminence.
n   Acetabulum.
o   Brim of acetabulum.
p   Notch of acetabulum.

q   Obturator foramen.
r   Horizontal branch of pubes.
s   Spine of pubes.
t   Descending ramus of pubes.
u   Symphysis pubis.
v   Ascending ramus of ischium.
w   Tuber ischii.
x   Descending ramus of ischium.

Fig. 3.

True or Sternal Ribs.

A.   Posterior extremity.
B.   Body.
C.   Anterior extremity.
a   Head.
b   Neck.
c   Tubercle.
d   Angle.
e   Inner or concave surface.
f   Outer or convex surface.
g   Outer or convex margin.
h   Inner or concave margin.
i   Inferior intercostal groove.

Fig. 4.

Sternum—anterior surface.

A.   Manubrium or first bone.
B.   Body or middle portion.
C.   Ensiform or xiphoid process.
a   Superior semilunar notch.
b   Clavicular fossa.
c   Articular fossa for first rib.
d   Articular fossa for second rib.
e   Articular fossa for true ribs.
f   Lateral semilunar notches.
g   Linea transversæ.
h   Supra-sternal osniculum.

Fig. 5.

Os Innominatum of right side—inner surface and lines of articulation of ilium, ischium, and pubes.

A.   Os ilium.
B.   Os ischii.

C.   Os pubis.
D.   Obturator foramen.
a   Crest of ilium.
b   Anterior superior spine of ilium.
c   Anterior inferior spine of ilium.
d   Posterior superior spine of ilium.
e   Posterior inferior spine of ilium.
f   Anterior semilunar notch.
g   Posterior semilunar notch.
h   Tuberosity of ilium.
i   Articular surface (sacro-iliac).
k   Internal semicircular ridge.
l   Great sciatic notch.
m   Body of ischium.
n   Descending ramus of ischium.
o   Ascending ramus of ischium.
p   Spine of ischium.
q   Tuber ischii.
r   Lesser sciatic notch.
s   Horizontal ramus of pubes.
t   Descending ramus of pubes.
u   Ilio-pectineal eminence.
v   Crest of pubes.
w   Spine of pubes.
x   Symphysis pubis.
y   Iliac fossa.

Fig. 6.

Os Innominatum of left side—outer surface.

A.—D. as in Fig. 5.
E.   Acetabulum.
a—w as in Fig. 5.
α   Brim of acetabulum.
β   Facies lunata (2 cornua).
γ   Fossa for ligamentum teres.
δ   Acetabular notch.

Figs. 7 & 8.

Os Coccygis—posterior surface (7); anterior and upper surfaces (8).

a—d   False vertebræ.
e   Articular surface.
f   Cornua coccygea.
g   False transverse processes.

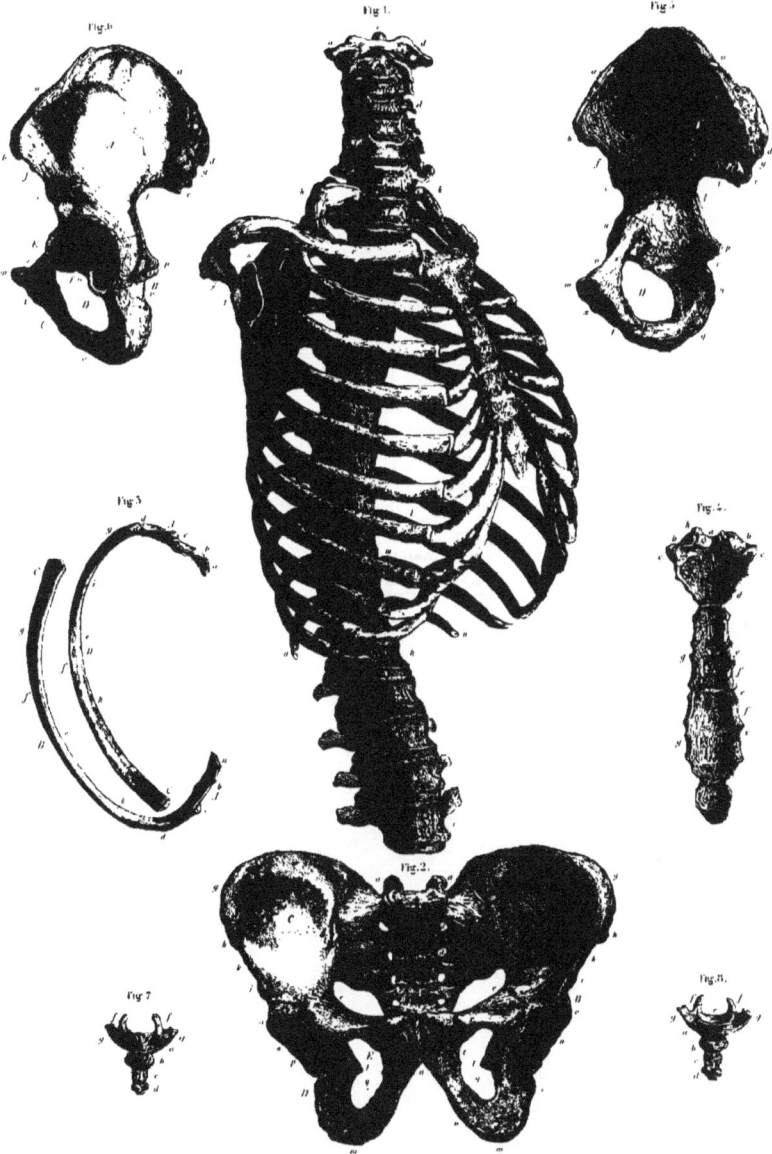

Tab. III.

Fig. 1.

Fig. 5.

Fig. 6.

Fig. 3.

Fig. 4.

Fig. 2.

Fig. 7.

Fig. 8.

PLATE IV.

BONES OF TRUNK (continued).

Fig. 1.

Posterior view of Trunk.

a Atlas. (*Vide* Fig. 2, 3, 4, and 5.)
b Dentata.
c Last cervical vertebra.
d First dorsal vertebra.
e Last dorsal vertebra.
f First lumbar vertebra.
g Last lumbar vertebra.
h Spinous processes.
i Transverse processes.
k Intervertebral foramina.
l First rib.
m Last rib.
n Clavicle.
o Scapula. (*Vide* Table V., Fig. 3.)
p Os sacrum.
q Os coccygis.
r Os ilium.
s Os ischii.
t Os pubis.
u Opening to sacral canal.
v Superior oblique processes of sacrum.
w False spinous processes.
x Posterior sacral foramina.
z Cornua sacralia.
a Termination of sacral canal.
β Cornua coccygea.
γ Crest of ilium.
δ Posterior superior spine of ilium.
ε Posterior semilunar notch of ilium.
ζ Posterior inferior spine of ilium.
η Great sciatic notch.
ϑ Superior semicircular ridge of ilium.
ι Inferior semicircular ridge of ilium.
κ Acetabulum.
λ Descending ramus of ischium.

μ Tuber ischii.
ν Spine of ischium.
ξ Lesser sciatic notch.
ϖ Obturator foramen.
ρ Ascending ramus of ischium.
ς Descending ramus of pubes.
  (*Vide* Table III., Fig. 1, 2, 5, 6, 7, and 8.)

Figs. 2 & 3.

Atlas (first Vertebra), Dentata (second Vertebra)—anterior surface (Fig. 2); posterior surface (Fig. 3.)

a Atlas.
b Dentata.
c Odontoid process.
d Articular surface of atlas for occipital condyle.

Figs. 4 & 5.

Atlas or first Cervical Vertebra—superior surface (Fig. 4); inferior surface (Fig. 5.)

a Anterior half arch.
b Posterior half arch.
c Lateral mass.
d Posterior tubercle of atlas.
e Articular surface for odontoid process.
f Condyloid fossa.
g Transverse process of atlas.
h Vertebral foramen.
i Groove for vertebral artery.
k Internal tubercle for transverse ligament.
l Spinal canal.
m Anterior tubercle of atlas.
n Inferior articular or oblique processes.

Fig. 6.

Dentata or second Cervical Vertebra—anterior surface.

a Body.
b Odontoid process.
c Neck of odontoid process.
d Articular surface for anterior half arch of atlas.
e Apex of odontoid process.
f Superior oblique processes.
g Inferior oblique processes.
h Transverse processes.

Fig. 7.

Cervical Vertebra—superior surface.

a Body.
b Arch.
c Spinous process.
d Interspinous cleft.
e Transverse processes.
f Superior oblique processes.
g Vertebral foramen.
h Spinal canal.

Figs. 8 & 9.

A Dorsal (Fig. 8) and a Lumbar Vertebra (Fig. 9)—superior surfaces.

a Body.
b Arch.
c Vertebral notch for intervertebral foramen.
d Spinous process.
e Transverse process.
f Articular surface for costal tubercle.
g Superior oblique processes.

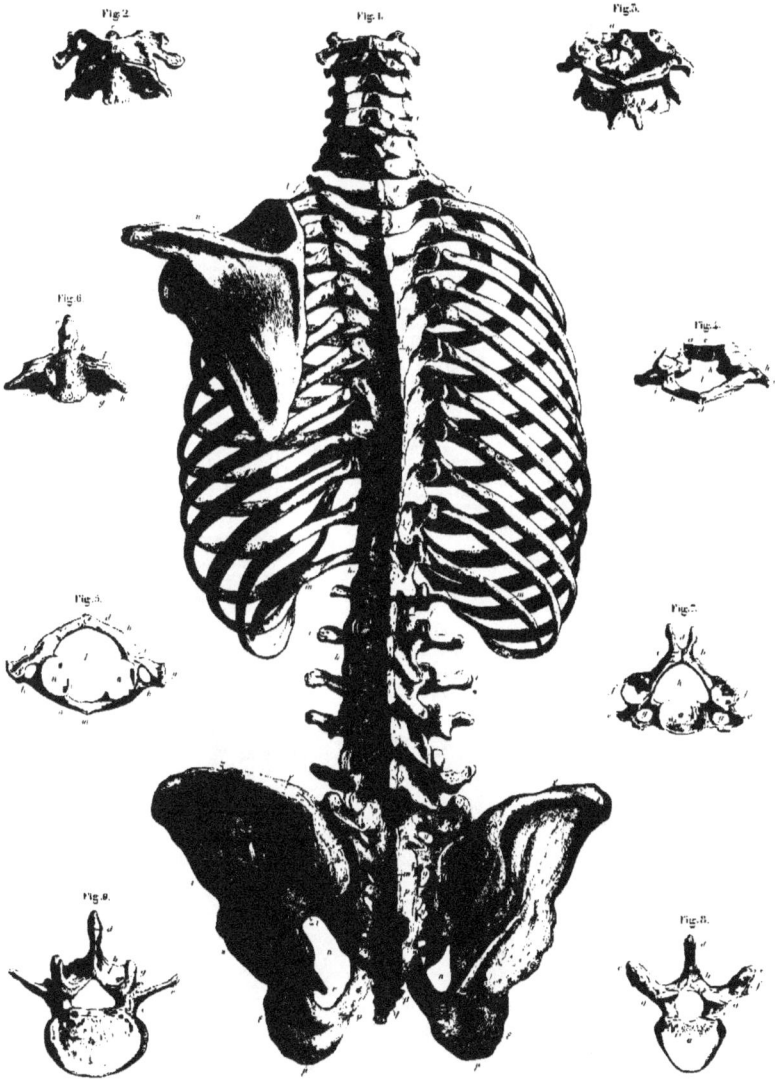

Tab. IV.

Fig. 2.   Fig. 1.   Fig. 3.

Fig. 6.   Fig. 4.

Fig. 5.   Fig. 7.

Fig. 9.   Fig. 8.

PLATE V.

BONES OF UPPER EXTREMITIES.

Fig. 1.

Clavicle (left)—superior surface.

a Body.
b Sternal end.
c Acromial end.
d Acromial articular surface.

Fig. 2.

Clavicle (left)—inferior surface.

a Body.
b Sternal end with articular surface.
c Acromial end.
d Tuberosity for costo-clavicular ligament.
e Tuberosity for corraco-clavicular ligament.

Fig. 3.

Scapula—posterior and outer surface.

a Supra-spinatus fossa.
b Infra-spinatus fossa.
c Spine.
d Acromion process.
e Articular surface for clavicle.
f Coracoid process.
g Supra-scapular notch.
h Superior margin.
i Superior posterior angle.
j Base or posterior margin.
k Inferior angle.
m Anterior inferior margin.
n Anterior angle or condyle.
o Neck.
p Glenoid cavity.

Fig. 4.

Scapula—anterior, internal, or concave surface.

a Subscapular fossa.
b Anterior angle or condyle.
c Glenoid cavity.
d Margin or brim of glenoid cavity.
e Acromion process.
f Articular surface for acromial end of clavicle.
g Coracoid process.
h Supra-scapular notch.
i Superior margin.
j Superior angle.
l Base or posterior margin.
m Inferior angle.
n Anterior margin.
o Tubercle for origin of triceps muscle.

Fig. 5.

Scapula—front view of anterior margin.

a Glenoid cavity.
b Brim of cavity.
c Anterior margin.
d Inferior angle.
e Spine.
f Acromion.
g Supra-spinatus fossa.
h Superior angle.
i Coracoid process.
k Tuberosity for long head of biceps.
l Subscapular fossa.

Fig. 6.

Humerus (left)—posterior view.

a Head of humerus.
b Greater tuberosity.
c Neck (anatomical).
d Body.
e External ridge.
f Internal ridge.
g Internal condyle.
h External condyle.
i Trochlea.
k Posterior olecranon fossa.

Fig. 7.

Humerus (left)—anterior view.

a Head of humerus.
b Neck (anatomical).
c Greater tuberosity.
d Lesser tuberosity.
e Bicipital groove.
f Inner margin of bicipital groove.
g Outer margin of bicipital groove.
h Body.
i Internal ridge.
k External ridge.
l Internal condyle.
m External condyle.
n Cubital process or epitrochlea.
o Rotula.
p Trochlea.
q Anterior humeral fossa.
r Anterior humeral fossa (lesser).

Fig. 8.

Ulna—posterior view.

a Olecranon process.
b Coronoid process.
c Greater sigmoid notch.
d Body.
e Internal angle.
f External angle.
g Capitulum.
h Styloid process.

Fig. 9.

Ulna—anterior view.

a Olecranon.
b Coronoid process.
c Greater sigmoid fossa.
d Lesser sigmoid fossa.
e Body.
f Capitulum.
g Styloid process.

Fig. 10.

Radius—anterior view.

a Capitulum.
b Glenoid cavity.
c Articular border.
d Neck.
e Tuberosity.
f Body.
g Inferior extremity.
h Styloid process.
i Ulnar fossa.
k Inferior glenoid cavity.
l Inner margin or crest.

Fig. 11.

Radius—posterior view.

a Capitulum.
b Articular border.
c Neck.
d Body.
e Crest.
f Inferior extremity.
g Styloid process.
h Ulnar fossa.
i Glenoid cavity.

Fig. 12.

Bones of Hand (right)—posterior or dorsal surface.

A. Carpus.
B. Metacarpus.
C. Fingers—phalanges.
a Os naviculare.
b Os lunare.
c Os triquetrum or cuneiforme.
d Os trapezium.
e Os trapezoides.

f Os magnum.
g Os unciforme.
i Metacarpal bone of thumb.
j Metacarpal bone of index finger (second).
k Metacarpal bone of middle finger (third).
l Metacarpal bone of ring finger (fourth).
m Metacarpal bone of little finger (fifth).
n Bases of metacarpal bones.
o Heads of metacarpal bones.
p First phalanx of thumb.
q Second phalanx of thumb.
r First phalanges of fingers.
s Second phalanges of fingers.
t Third or ungual phalanges of fingers.

Fig. 13.

Carpus, Metacarpus, and Phalanges of Thumb (left)—posterior surface.

a Os naviculare.
b Os lunare.
c Os triquetrum or cuneiforme.
d Os pisiforme.
e Os trapezium.
f Os trapezoides.
g Os magnum.
h Os unciforme.
i Metacarpal bones.
k Bases of metacarpal bones.
l Heads of metacarpal bones.
m Sesamoid bones.

Fig. 14.

Carpus, Metacarpus, and Phalanges of Thumb (left)—anterior surface.

a–h as in Fig. 13.
k Hamular process of unciform bone.
k Metacarpal bones.
l Bases of metacarpal bones.
m Heads of metacarpal bones.

Fig. 15.

Carpal Bones, (left) first row—superior or convex articular surface.

a Os naviculare.
b Os lunare.
c Os triquetrum.
d Os pisiforme.
e Os triquetrum.
f Os unciforme.
g Eminence of radial carpus (1).
h Eminence of radial carpus (2).
i Eminence of ulnar carpus (1).
k Eminence of ulnar carpus (2).

Fig. 16.

Carpal Bones, (left) second row—intercarpal articular surface.

a Os trapezium.
b Os trapezoides.
c Os magnum.
d Os unciforme.
e Hamular process of unciform bone.

Fig. 17.

Carpal bones, (left) first row—inferior articular surface.

a Os naviculare.
b Os lunare.
c Os triquetrum.
d Os pisiforme.

Fig. 18.

Carpal bones, (left) second row—digital surface.

a Os trapezium.
b Os trapezoides.
c Os magnum.
d Os unciforme.
e Hamular process of unciform bone.
f Os pisiforme.

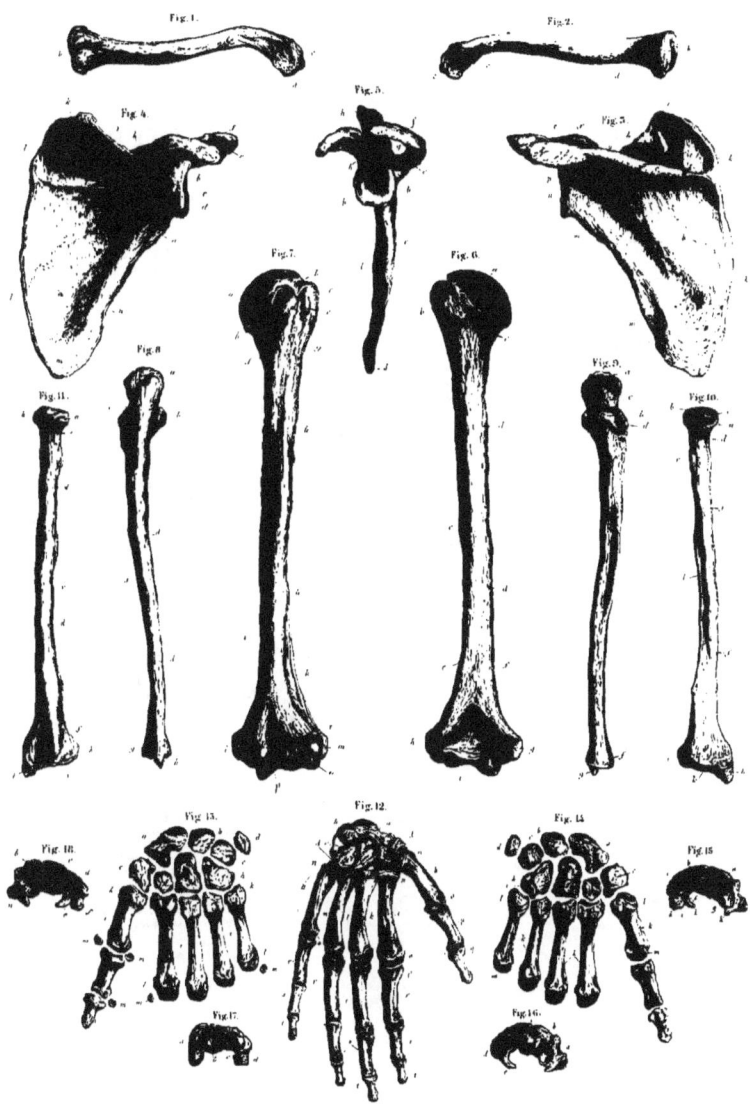

Fig. 1.

Fig. 2.

Fig. 3.

Fig. 4.

Fig. 5.

Fig. 6.

Fig. 7.

Fig. 8.

Fig. 9.

Fig. 10.

Fig. 11.

Fig. 12.

Fig. 13.

Fig. 14.

Fig. 15.

Fig. 16.

Fig. 17.

Fig. 18.

PLATE VI.

BONES OF LOWER EXTREMITIES.

Fig. 1.

Os Femoris, Femur (left)—anterior surface.

a Head.
b Fossa for ligamentum teres.
c Neck.
d Trochanter major.
e Trochanter minor.
f Anterior inter-trochanteric line.
g Body.
h External condyle.
i Internal condyle.
k Articular surface for patella.

Fig. 2.

Os Femoris, Femur (left)—posterior surface.

a—e as in Fig. 1:
f Posterior inter-trochanteric line.
g Superior oblique lines of linea aspera.
h Linea aspera.
i Inferior oblique line of linea aspera.
k Body.
l Popliteal fossa.
m External condyle.
n Internal condyle.
o Inter-condyloid fossa.

Fig. 3.

Patella (left)—anterior surface.

a Base.
b Body.
c Apex.
d Internal margin.
e External margin.

Fig. 4.

Patella (left)—posterior or articular surface.

a Base.
b Apex.
c Internal margin.
d External margin.
e Articular surface.

Fig. 5.

Tibia (left)—anterior and inner surface.

a Internal condyle.
b External condyle.
c Internal articular surface.
d External articular surface.

e Inter-condyloid eminence.
f Articular surface for head of fibula.
g Tubercle.
h Body.
i Spine or crest.
k Fossa for external malleolus.
l Inferior articular surface for astragalus.
m Internal malleolus.
n Inner surface.
o Outer surface.

Fig. 6.

Tibia (left)—posterior surface.

a—f as in Fig. 5.
g Oblique line.
h Body.
i Internal malleolus.
k Fibular fossa.
l Articular surface for astragalus.

Fig. 7.

Fibula (left)—anterior surface.

a Capitulum or head.
b Superior articular surface.
c Body.
d External malleolus.
e Tibial surface.
f Astragalean articular surface.

Fig. 8.

Fibula (left)—posterior surface.

a—f as in Fig. 7.

Fig. 9.

Bones of Foot (right)—upper or dorsal surface.

a Astragalus.
b Os calcis.
c Os naviculare.
d Internal cuneiform.
e Middle cuneiform.
f External cuneiform.
g Os cuboides.
h First metatarsal.
i Metatarsal bones.
k First phalanx of great toe.
l Second phalanx of great toe.
m First phalanges.
n Second phalanges.
o Third or ungual phalanges.

Fig. 10.

Bones of Foot (right)—inferior or plantar surface.

a—o as in Fig. 9.

Fig. 11.

Tarsal and Metatarsal Bones (left)—upper or dorsal surface.

I. Astragalus.
  a Body.
  b Neck.
  c Head.
  d Superior tibial articular surface.
II. Os calcis.
  e Body.
  f Tuberosity.
  g Anterior process.
  h Superior articular surface for astragalus.
  i Sustentaculum.
  k Lateral articular surface for astragalus.
  l Anterior articular surface for os naviculare.
III. Os naviculare.
  m Tuberosity for tendon of tibialis posticus.
IV. Internal cuneiform bone.
V. Middle cuneiform bone.
VI. External cuneiform bone.
VII. Os cuboides.
  n Metatarsal bones (body).
  o Bases of metatarsal bones.
  p Capitals of metatarsal bones.
  q Tuberosity of fifth metatarsal bone.
  r Sesamoid bones of great toe.

Fig. 12.

Tarsal and Metatarsal Bones (left)—under or plantar surface.

I.—VII. as in Fig. 11.
a—c as in Fig. 11.
d Inferior articular surface for body of os calcis.
e Anterior articular surface for interior process of os calcis.
f Astragalean groove.
g Body of os calcis.
h Tuberosity of os calcis.
i Anterior process of os calcis.
k Anterior articular surface of os calcis.
l Sustentaculum of os calcis.
m Cuboid groove for peroneus longus tendon.
n—r as in Fig. 11.
s Plantar tubercle of first metatarsal bone.
t Tuber of os naviculare.

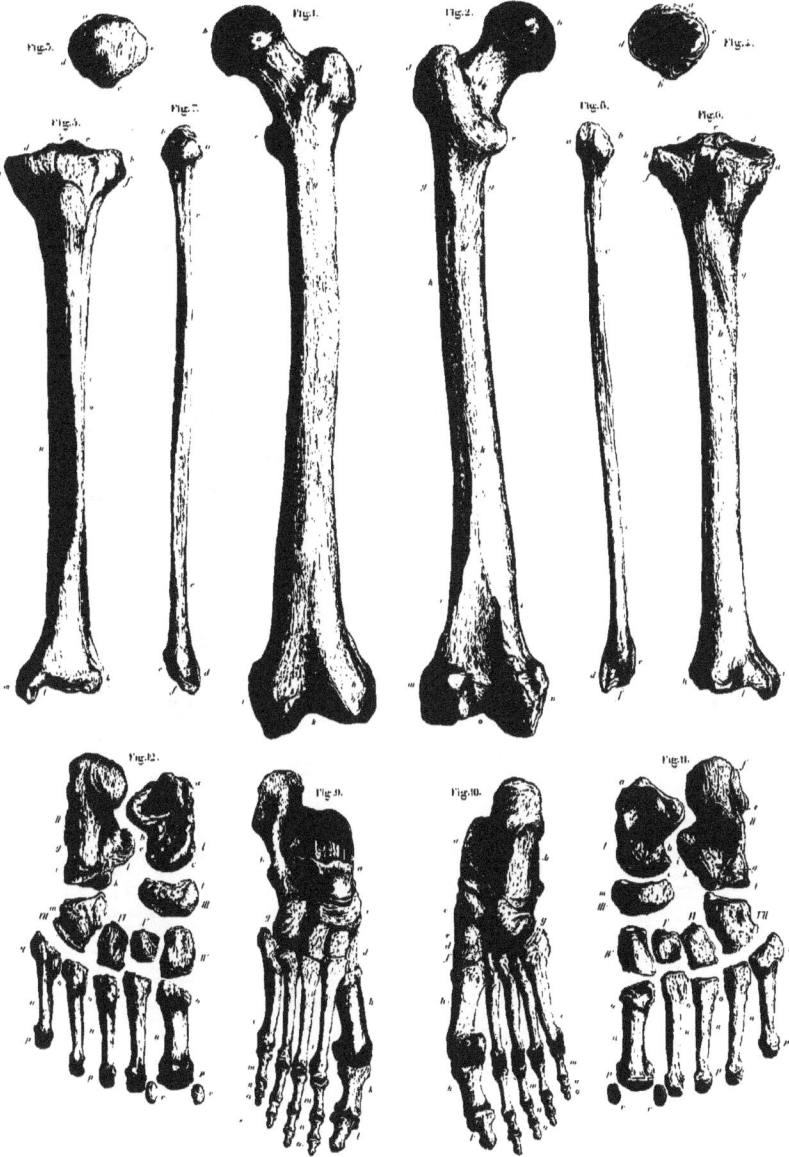

Tab.VI.

PLATE VII.

LIGAMENTS OF HEAD, TRUNK, AND UPPER EXTREMITIES.

Fig. 1.

Ligaments of the Vertebræ, Sternal end of Ribs, Pelvis, and Ilio-femoral articulation—anterior surface.

a   Body of sphenoid bone.
b   Temporal bones.
c   Atlas or first vertebra.
d   Transverse process of atlas.
e   Dentata or second vertebra.
f   Seventh cervical vertebra.
g   First dorsal vertebra.
h   Last dorsal vertebra.
i   Spinal end of first rib.
k   Spinal end of twelfth rib.
l   First lumbar vertebra.
m   Last lumbar vertebra.
n   Os sacrum.
o   Os coccygis.
p   Os ilium.
q   Crest of ilium.
r   Anterior superior spine of ilium.
s   Anterior inferior spine of ilium.
t   Horizontal ramus of pubes.
u   Descending ramus of pubes.
v   Symphysis pubis.
w   Ascending ramus of ischium.
x   Tuber ischii.
y   Descending ramus of ischium.
(For bones of pelvis vide Plate III.)
  1. Anterior intervertebral ligament.
  2. Anterior occipito-atlantoid ligament.
  3. Intervertebral fibro-cartilage.
  4. Intertransverse ligaments.
  5. Posterior costo-vertebral ligaments.
  6. Internal costo-transverse ligaments.
  7. External costo-transverse ligaments.
  8. Posterior intercostal ligaments.
  9. Lumbo-costal ligaments.
 10. Superior ilio-lumbar ligaments.
 11. Inferior ilio-lumbar ligaments.
 12. Anterior ilio-sacral ligaments.
 13. Lesser sciatic ligaments.
 14. Anterior sacro-coccygeal ligament.
 15. Obturator ligaments.
 16. } Capsular ligament of hip.
 17. }
 18. Accessory ligaments of hip.
 19. Bursa mucosa of iliacus internus muscle.
 20. Sub-pubic ligament.
 21. Inter-pubic ligament.

Fig. 2.

Ligaments of right Temporo-maxillary articulation—external surface.

a   Temporal bone.
b   Meatus auditorius externus.
c   Zygomatic arch.
d   Ramus of inferior maxilla.
e   Condyloid process of maxilla.
f   Coronoid process of maxilla.
g   Angle of maxilla.
h   Styloid process.
  1. Capsular ligament.
  2. External lateral ligament.
  3. Stylo-maxillary ligament.
  4. Stylo-hyoid ligament.

Fig. 3.

Ligaments of right Temporo-maxillary articulation—internal surface.

a   Temporal bone.
b   Sphenoid bone.
c   Pterygoid process.

d   Ramus of inferior maxilla.
e   Condyloid process of maxilla.
f   Angle of maxilla.
  1. Capsular ligament.
  2. Internal lateral ligament.

Figs. 4 & 5.

Internal Ligaments connecting Occipital Bone with Dentata and of the articulation between Atlas and Dentata—posterior view, the posterior half arches of these bones having been removed.

a   Os occipitis.
b   Atlas.
c   Dentata.
d   Odontoid process.
  1. Posterior longitudinal ligament.
  2. Apparatus ligamentosus.
  3. Crucial ligament.
 4. 4. Transverse portion } of crucial liga-
  5. Superior appendix } ment.
  6. Inferior appendix }
 7. 7. Lateral inferior ligaments of odontoid process.
 8. 8. Lateral superior ligaments of odontoid process.
  9. Capsular ligament between atlas and occipital bone.
 10. Lateral occipito-transverse ligament.
 11. Capsular ligament between atlas and dentata.
 12. Suspensory ligament of odontoid process.

Figs. 6 & 7.

Ligaments of Sterno-clavicular and Sterno-costal articulations with anterior Intercostal Ligaments—anterior surface (Fig. 6); posterior surface (Fig. 7).

a   Sternal end of clavicle.
b   First bone of sternum.
c   Body of sternum.
d   Ensiform process.
e   First rib.
f   Seventh rib.
g   First false or eighth rib.
  1. Interclavicular ligament.
  2. Internal capsular ligament of sterno-clavicular articulation.
  3. Rhomboid ligament.
 4. 4. Ligamenta coruscantia.
  5. Anterior proper sternal ligament.
  6. Posterior proper sternal ligament.
 7. 7. Radiated ligaments of costal cartilages (anterior and posterior).
  8. Costo-xiphoid ligaments.

Figs. 8 & 9.

Ligaments of Shoulder-joint and scapulo-clavicular articulation.

a   Acromial end of clavicle.
b   Acromion.
c   Scapula.
d   Coracoid process.
e   Glenoid cavity.
f   Spine of scapula.
g   Os humeri.
h   Head of os humeri.
i   Greater tuberosity of humerus.
  1. Claviculo-acromial ligament.
  2. External capsular ligament of clavicle.
  3. Trapezoid ligament } or coraco-clavi-
  4. Conoid ligament. } cular ligament.

  5. Coraco-acromial ligament.
  6. Transverse ligament of scapula.
  7. Capsular ligament of shoulder-joint.
  8. Tendon of long head of biceps.
  9. Glenoid ligament.

Figs. 10 & 11.

Ligaments of left Elbow-joint—anterior left surface (Fig. 10); posterior surface (Fig. 11).

a   Os humeri.
b   External condyle.
c   Internal condyle.
d   Trochlea.
e   Radius.
f   Neck
g   Head } of radius.
h   Tubercle }
i   Ulna.
k   Olecranon.
  1. Capsular ligament.
  2. External lateral ligament.
  3. Internal lateral ligament.
  4. Orbicular ligament of radius.
  5. Oblique ligament of radio-ulnar articulation.
  6. Interosseous ligament.

Fig. 12.

Ligaments of Left Wrist-joint and Hand.

a   Lower end of radius.
b   Lower end of ulna.
c   Styloid process of radius.
d   Styloid process of ulna.
e   Os naviculare.
f   Os lunare.
g   Os triquetrum.
g   Os trapezium.
h   Os trapezoides.
i   Os magnum.
k   Os uniforme.
  1. Interosseous ligament.
  2. Internal lateral ligament.
  3. External lateral ligament.
  4. Posterior radio-carpal ligament.
  5. Posterior superficial carpal ligaments.
  6. Posterior deep carpal ligaments.
  7. Internal lateral ligament of carpus.
  8. Proper ligaments of carpus.
  9. Dorsal carpo-metacarpal ligaments.
 10. 10. Dorsal ligaments of metacarpal bones.
 11. 11. External lateral ligaments of fingers.
 12. Internal lateral ligaments of fingers.

Fig. 13.

Ligaments of Left Wrist-joint and Hand—anterior surface.

a—l as in Fig. 13.
m   Hamular process.
n   Os pisiforme.
  1. Interosseous ligament.
 2—3. Anterior radio-carpal ligaments.
  4. Lateral radial ligament.
  5. Lateral ulnar ligament.
  6. Triangular cartilage.
 7. 7. Anterior proper carpal ligaments.
 8. 8. Anterior carpo-metacarpal ligaments.
 9. 9. Anterior inter-metacarpal ligament.
 10. 10. Anterior ligament of metacarpo-phalangeal articulation.
 11. 11. External lateral ligaments of metacarpo-phalangeal articulation.
 12. Internal lateral ligaments of metacarpo-phalangeal articulation.

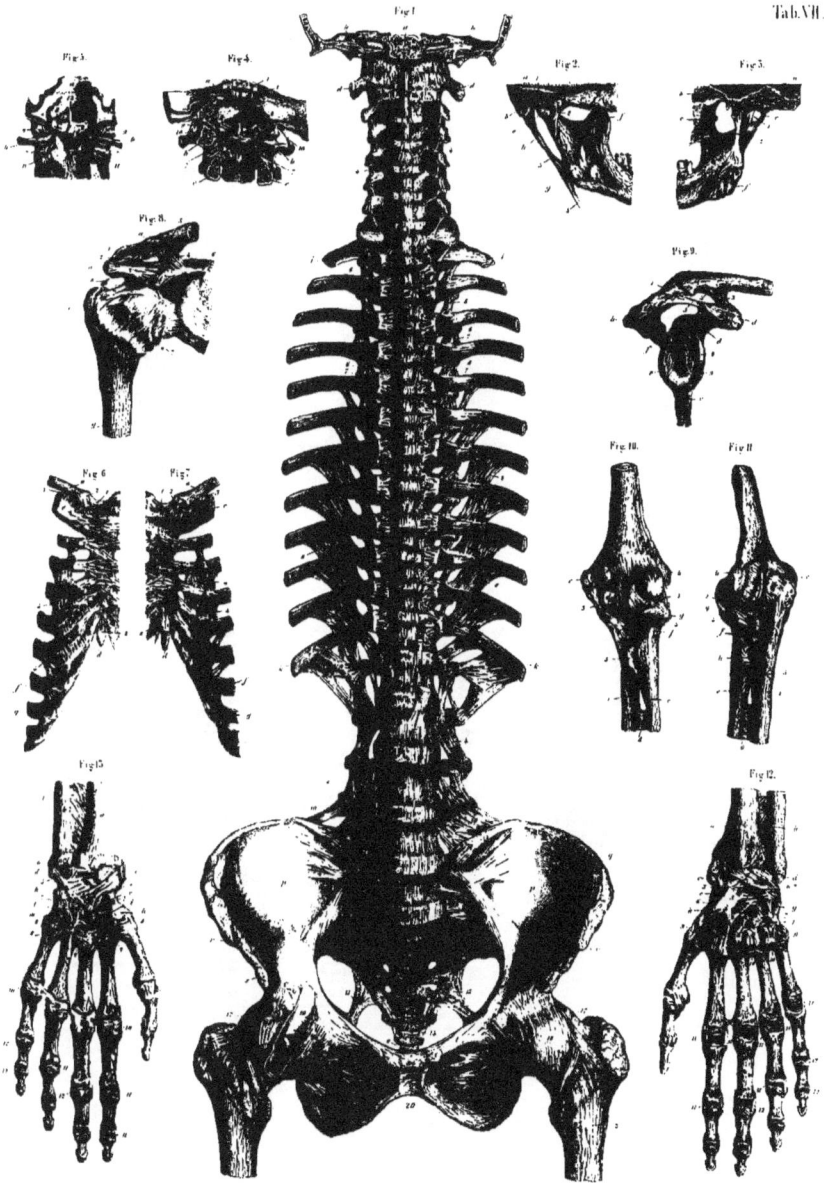

Tab.VII.

PLATE VIII.

LIGAMENTS OF SPINE, PELVIS, AND JOINTS OF INFERIOR EXTREMITIES.

Fig. 1.

Ligaments of Cervical Vertebræ—posterior surface, and anterior of spinal canal, of dorsal spines, and costo-vertebral articulations.

a   Os occipitis.
b   Mastoid process.
c   Fossa for medulla oblongata.
d   Atlas
e   Transverse process of atlas.
f   Dentata.
g—l   Third to seventh cervical vertebra.
m—p   F rst to fourth dorsal vertebra.
q—t   First to fourth rib.
   1. Superior attachment of ligamentum posticum.
   2. Apparatus ligamentosus colli.
   3. Capsular ligament between occipital bone and atlas.
   4. Interspinal ligament.
   5. Posterior costo-transverse ligament.
   6. Internal and external ligaments of necks of ribs.

Fig. 2.

Posterior or Dorsal Ligaments of Spinal Column, Pelvis, and Ilio-femoral articulations.

a—c Tenth to last dorsal vertebra.
d—h First to last lumbar vertebra.
i—l Tenth to last rib.
m   Os sacrum.
n   Sacral opening of spinal canal.
o   Os coccygis.
p   Os ilium.
q   Os ischii.
r   Os pubis.
s   Os femoris.
(For bones of pelvis vide Plate IV.)
   1. Inter-spinous ligaments.
   2. Posterior intercostal ligaments.
   3. Lumbo-costal ligaments.
   4. External transverse ligaments.
   5. Internal transverse ligaments.
   6. Superior ilio-lumbar ligament.
   7. Inferior ilio-lumbar ligament.
   8. Long ilio-sacral ligament.
   9. Short ilio-sacral ligament.
   10. Inferior ilio-sacral ligament.
   11. Posterior irregular ligaments.
   12. Posterior sacro-coccygeal ligaments.
   13. Greater sacro-sciatic ligament.
   14. Lesser sacro-sciatic ligament.
   15. Obturator ligament.
   16. Sub-pubic ligament.
   17. Capsular ligament of hip-joint.
   18. Transverse fibres of capsular ligament.
   19. Vertical fibres of capsular ligament
   20. Falciform portion of great sacro-sciatic ligament.

Fig. 3.

Ligaments of Left Knee-joint—internal and anterior surface.

a   Os femoris.
b   Internal condyle of femur.
c   Patella.
d   Tibia.
e   Tuberosity of tibia.
f   Tendon of extensor communis.
   1. Ligamentum patellæ.
   2. Internal fibres of ligamentum patellæ.
   3. Internal lateral ligament.
   4. Capsular ligament.

Fig. 4.

Ligaments of Left Knee-joint—internal anterior view.

a   Os femoris.
b   Internal condyle of femur.
c   External condyle of femur.
d   Articular surface for patella.
e   Tibia.
f   Internal condyle of tibia.
g   External condyle of tibia.
h   Tuberosity of tibia.
i   Fibula.
k   Head of fibula.
l   Inter-condyloid eminence.
   1. External semilunar cartilage.
   2. Internal semilunar cartilage.
   3. Anterior crucial ligament.
   4. Posterior crucial ligament.
   5. Internal lateral ligament.
   6. Capsular ligament of head of fibula.
   7. Interosseous membrane of leg.

Fig. 5.

Ligaments of Left Knee-joint—posterior surface.

a   Os femoris.
b   Internal condyle of femur.
c   External condyle of femur.
d   Inter-condyloid fossa.
e   Tibia.
f   External condyle of tibia.
g   Internal condyle of tibia.
h   Fibula.
1—7 as in Fig. 4.

Fig. 6.

Ligaments of the Sole or Plantar surface of Left Foot.

a   Corpus of os calcis.
b   Tuberosity of os calcis.
c   Anterior process of os calcis.
d   Astragalus.
e   Os naviculare.
f   Tuber of os naviculare.
g   Internal cuneiform bone.
h   Middle cuneiform bone.
i   External cuneiform bone.
k   Os cuboides.
l   Plantar tubercle of first metacarpal bone.
m   Tuberosity of fifth metatarsal bone.
n   Metatarsal bones.
o   Lesser heads of metatarsal bones.
   1. Astragalo-calcanean ligaments.
   2. Calcaneo-cuboid or ligamentum longum.
   3. Calcaneo-navicular ligament.
   4. Cuboideo-navicular ligament.
   5. Naviculo-cuneiform ligament.
   6. Cuboideo-cuneiform ligament.
   7. Inter-cuneiform ligaments.
   8. Cuboideo-metatarsal ligament of fifth toe.
   9. Cuneiformo-metatarsal ligament of great toe.
   10. Ligaments of bases of metatarsal bones.
   11. Cuboideo-metatarsal ligaments.
   12. Ligaments of lesser heads of metatarsal bones.
   13. Fibro-cartilaginous sheaths for flexor tendons.
   14. Internal lateral ligaments of phalanges.
   15. External lateral ligaments of phalanges.
   16. Crucial ligaments.
   17. Inter sesamoid ligaments.

Fig. 7.

Ligaments of Left Foot—internal surface.

a   Tibia.
b   Fibula.
c   Malleolus internus.
d   Astragalus.
e   Os calcis.
f   Tuberosity of os calcis.
g   Sustentaculum.
h   Os cuboides.
i   Os naviculare.
k   Internal cuneiform bone.
l   Middle cuneiform bone.
m   Metatarsal bone of great toe.
n   Base of first metatarsal bone.
o   Lesser head of first metatarsal bone.
p   First phalanx of great toe.
q   Second phalanx of great toe.
r   Second toe.
   1. Internal lateral or deltoid ligament.
   2. Posterior ligament of ankle.
   3. Posterior astragalo-calcanean ligament.
   4. Plantar calcaneo-cuboid ligament.
   5. Dorsal astragalo-navicular ligament.
   6. Calcaneo-navicular ligament.
   7. Lateral naviculo-cuneiform ligament.
   8. Dorsal naviculo-cuneiform ligament (1).
   9. Dorsal naviculo-cuneiform ligament (2).
   10. Dorsal inter-cuneiform ligament.
   11. Dorsal ligament of base of first metatarsal bone.
   12. Plantar ligament.
   13. Internal lateral ligaments of toes.

Fig. 8.

Ligaments of Left Foot—external and dorsal surfaces.

a   Fibula.
b   Tibia.
c   Malleolus externus.
d   Corpus of os calcis.
e   Tuber of os calcis.
f   Process of os calcis.
g   Os cuboides.
i   Astragalus.
i   Os naviculare.
k   Middle cuneiform bone.
l   External cuneiform bone.
m   Internal cuneiform bone.
n   Tubercle of fifth metatarsal bone.
   1. Interosseous membrane of leg.
   2. Posterior tibio-fibular ligament.
   3. Anterior superior tibio-fibular ligament.
   4. Anterior inferior tibio-fibular ligament.
   5. External lateral ligament of ankle, posterior and middle portions.
   6. Posterior portion of external lateral ligament of ankle.
   7. Anterior portion of external lateral ligament of ankle.
   8. Tarsal apparatus ligamentosus.
   9. Dorsal calcaneo-cuboid ligament.
   10. Plantar calcaneo-cuboid ligament.
   11. Dorsal calcaneo-navicular ligament.
   12. Dorsal astragalo-navicular ligament.
   13. Dorsal cuboideo-navicular ligament.
   14. Dorsal naviculo-cuneiform ligament (3).
   15. Dorsal naviculo-cuneiform ligament (2).
   16. Dorsal inter-cuneiform ligaments.
   17. Dorsal cuboideo-metatarsal ligament of fifth toe.
   18. Dorsal interosseous ligaments of tarsus and metatarsus.
   19. Dorsal ligaments of bases of metatarsal bones.
   20. External lateral ligaments of toes.

Fig. 3.

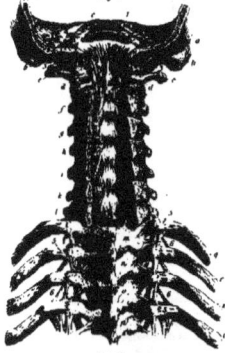

Fig. 1.

Fig. 5.

Fig. 6.

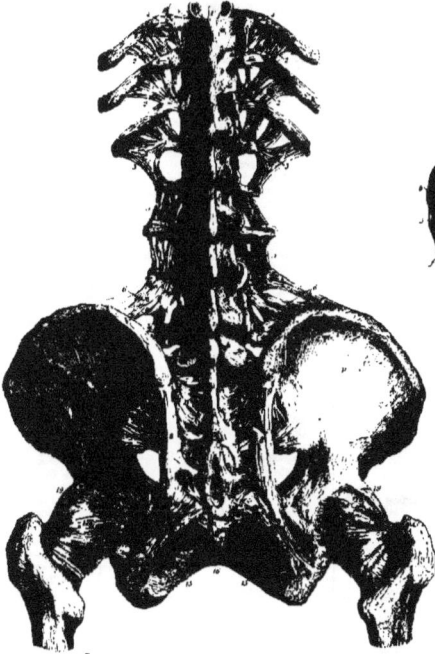

Fig. 2.

Fig. 4.

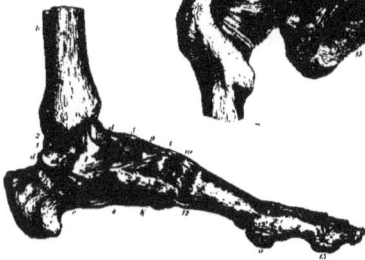

Fig. 7.

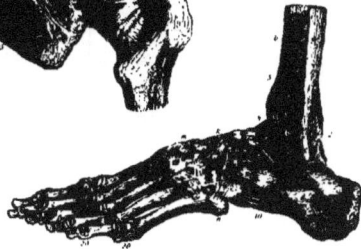

Fig. 8.

PLATE IX.

MUSCLES OF HEAD AND NECK.

Fig. 1.

Muscles of Face and Neck—anterior surfaces.

a   Os frontis.
b   External angular process of frontal bone.
c   Malar bone.
d   Zygomatic process of malar bone.
e   Superior maxilla.
f   Inferior maxilla.
g   Ala nasi.
h   Clavicle.
i   Inferior lip.
k   Superior eyelid.
l   Inferior eyelid.
m   Internal palpebral ligament.
1.   Frontal portion of occipito-frontalis muscle.
2.   Tendon of occipito-frontalis muscle.
3.   Pyramidalis nasi.
4.   Orbicularis palpebrarum, upper portion.
5.   Orbicularis palpebrarum, lower portion.
6.   Corrugator supercilii.
7.   Levator labii superioris alæque nasi.
8.   Levator labii superioris proprius.
9.   Zygomaticus minor.
10.   Zygomaticus major.
11.   Levator anguli oris.
12.   Levator palpebræ superioris tendon.
13.   Buccinator.
14.   Orbicularis oris.
15.   Depressor anguli oris, or triangularis menti.
16.   Depressor labii inferioris, or quadratus menti.
17.   Levator menti.
18.   Masseter.
19.   Temporal.
20.   Sterno-cleido-mastoid.
21.   Clavicular origin of sterno-cleido-mastoid.
22.   Sternal origin of sterno-cleido-mastoid.
23.   Sterno-hyoid.
24.   Sterno-thyroid.
25.   Anterior margin of trapezius.
26.   Omo-hyoid (anterior belly with posterior belly and tendon).
27.   Levator anguli scapulæ.
28.   Scalenus anticus.
29.   Scalenus medius.
30.   Attrahens auris.
31.   Compressor naris.

Fig. 2.

Muscles of Neck—lateral surface of right side.

a   Corpus of inferior maxilla.
b   Angle of inferior maxilla.
c   Ramus of inferior maxilla.
d   Mastoid process.
e   Occipital bone.

f   Clavicle.
g   Scapula.
h   Acromion.
i   Coracoid process.
k   First rib.
l   Manubrium sterni.
m   Body of os hyoides.
n   Greater bone of os hyoides.
o   Styloid process.
p   Thyroid cartilage.
q   Thyroid gland.
r   Trachea.
s   Œsophagus.
1.   Anterior belly of digastric.
2.   Posterior belly of digastric.
3.   Hyo-digastric membrane.
4.   Mylo-hyoideus.
5.   Hyo-glossus.
6.   Stylo-hyoid.
7.   The perforation of the stylo-hyoid muscle by the digastric tendon.
8.   Stylo-glossus.
9.   Stylo-pharyngeus.
10.   Middle constrictor of pharynx.
11.   Inferior constrictor of pharynx.
12.   Thyro-hyoid membrane.
13.   Thyro-hyoid muscle.
14.   Sterno-hyoid.
15.   Sterno-thyroid.
16.   Anterior belly of omo-hyoid.
17.   Posterior belly of omo-hyoid.
18.   Tendon of omo-hyoid.
19.   Longus colli.
20.   Rectus capitis anticus major.
21.   Scalenus anticus.
22.   Scalenus medius et posticus.
23.   Levator anguli scapulæ.
24.   Splenius capitis.
25.   Sterno-cleido-mastoideus.
26.   Obliquus capitis superior.
27.   Obliquus capitis inferior.
28.   Trapezius v. cucullaris.
29.   Deltoideus.

Fig. 3.

Muscles of Neck—front view.

a   Inferior maxilla.
b   Os hyoides.
c   Larynx.
d   Thyroid gland.
e   Trachea.
f   Mastoid process.
g   Clavicle.
h   Manubrium sterni.
i   Lingualis muscle.
1.   Anterior belly of digastric.
2.   Posterior belly of digastric.
3.   Hyo-digastric membrane.
4.   Mylo-hyoideus.
5.   Hyo-glossus.

6.   Stylo-hyoideus.
7.   Stylo-glossus.
8.   Stylo-pharyngeus.
9.   Genio-hyoideus.
10.   Thyro-hyoideus.
11.   Sterno-thyroid.
12.   Inferior constrictor of pharynx.
13.   Sterno-hyoid.
14.   Posterior belly of omo-hyoid.
15.   Anterior belly of omo-hyoid.
16.   Crico-thyroideus.
17.   Longus colli.
18.   Scalenus anticus.
19.   Scalenus medius.
20.   Scalenus posticus.
21.   Levator anguli scapulæ.
22.   Splenius capitis.

Fig. 4.

Deep Muscles of right side of Neck.

a   Mastoid process.
b   Zygomatic arch.
c   Meatus auditorius externus.
d   Inferior maxilla.
e   Styloid process.
f   Transverse process of atlas.
g   Superior maxilla.
h   Os hyoides.
i   Larynx.
k   Trachea.
l   Clavicle.
m   First rib.
n   Acromion.
o   Coracoid process.
p   Œsophagus.
q   Spine of scapula.
1.   Orbicularis oris.
2.   Buccinator.
3.   Superior constrictor of pharynx.
4.   Stylo-glossus.
5.   Stylo-pharyngeus.
6.   Middle constrictor of pharynx.
7.   Hyo-glossus.
8.   Mylo-hyoid.
9.   Thyro-hyoid.
10.   Inferior constrictor of pharynx.
11.   Thyro-hyoid membrane.
12.   Crico-thyroid muscle.
13.   Rectus capitis anticus major.
14.   Scalenus anticus.
15.   Scalenus medius.
16.   Scalenus posticus.
17.   Levator anguli scapulæ.
18.   Splenius capitis.
19.   Serratus posticus superior.
20.   Rhomboideus superior.
21.   Trapezius v. cucullaris.
22.   Supraspinatus.
23.   Sterno-thyroid (upper portion thrown back).

Tab.IX.

Fig.1

Fig.2

Fig.4

Fig.5

PLATE X.

MUSCLES OF POSTERIOR PART OF NECK, TRUNK, PHARYNX, PALATE, INFERIOR MAXILLA, AND TONGUE.

Fig. 1.

Muscles of back of Pharynx and Inferior Maxilla.

a   Basilar process.
b   Petrous portion of temporal bone.
c   Ramus of inferior maxilla.
d   Os hyoides (posterior cornua).
o   Thyroid cartilage.
f   Tayro-hyoid ligament.
g   Œsophagus.
h   Trachea.
i   Styloid process.
k   Stylo-maxillary ligament.
   1. Superior constrictor of pharynx.
   2. Middle constrictor of pharynx.
   3. Inferior constrictor of pharynx.
   4. Stylo-pharyngeus.
   5. Stylo-glossus.
   6. Mylo-hyoideus.
   7. Pterygoideus internus.
   8. Masseter.
   9. Buccinator.

Fig. 2.

Muscles of Palate and Throat—posterior view, the pharyngeal constrictors being divided and thrown back. Internal and posterior view of Nares, Fauces, and Larynx.

a   Basilar process.
b   Petrous bone.
c   Ramus of inferior maxilla.
d   Styloid process.
e   Posterior nares.
f   Condyle of inferior maxilla.
g   Base of tongue.
h   Epiglottis.
i   Posterior and upper border of cricoid cartilage.
k   Œsophagus.
l   Trachea.
   1. Constrictor pharyngis superior.
   2. Constrictor pharyngis media.
   3. Constrictor pharyngis inferior.
   4. Azygos uvula.
   5. Levator palati mollis.
   6. Circumflexus palati mollis.
   7. Crico-arytænoideus posticus.
   8. Palato-pharyngeus.

Fig. 3.

Muscles of Tongue—lateral view of right side.

a   Corpus of inferior maxilla.
b   Ramus of inferior maxilla.
c   Styloid process.
d   Os hyoides.
e   Larynx.
f   Tongue.
   1. Lingualis.
   2. Genio-glossus.
   3. Hyo-glossus.
   4. Stylo-glossus.
   5. Stylo-pharyngeus.
   6. Genio-hyoideus.
   7. Mylo-hyoideus.
   8. Thyro-hyoid membrane.

Fig. 4.

Internal Muscles of Inferior Maxilla, with posterior view of Mouth and Nares.

a   Body of sphenoid bone.
b   Petrous bone.
c   Condyle of inferior maxilla.
d   Ramus of inferior maxilla.
o   Body of inferior maxilla.
f   Hard palate.
g   Pterygoid process.
h   Posterior nares.
   1. Pterygoideus internus.
   2. Pterygoideus externus.
   3. Masseter.
   4. Mylo-hyoideus (divided).
   5. Genio-glossus (divided).
   6. Genio-hyoideus (divided).

Fig. 5.

Muscles of Soft Palate—posterior and internal view.

a   Body of sphenoid bone.
b   Petrous bone.
c   Condyle of inferior maxilla.
d   Ramus of inferior maxilla.
e   Hard palate.
f   Pterygoid process.
g   Hamular process.
h   Posterior nares.
i   Eustachian tube.
   1. Pterygoideus externus.

2. Levator palati mollis.
3. Circumflexus palati mollis.
4. Azygos uvula.
5. Palato-pharyngeus.

Fig. 6.

Muscles of posterior surface of Neck and upper part of Thorax beneath the trapezius muscle.

a   Occipital bone.
b   Superior semicircular line.
c   Mastoid process.
d   Atlas.
e   Seventh cervical vertebra.
f   First dorsal vertebra.
g   Eighth dorsal vertebra.
h–l   First to fifth true rib.
m   Ligamentum nuchæ.
n   Ligamentum apicum.
   1. Splenius capitis.
   2. Splenius colli.
   3. Serratus posticus superior.
   4. Biventer cervicis.
   5. Complexus cervicis.
   6. Transversalis cervicis.
   7. Longissimus dorsi.

Fig. 7.

Deep Muscles of Neck and Back.

a   Os occipitis.
b   Mastoid process.
c   Spinous process of seventh cervical vertebra.
d   First rib.
e   Last rib.
f   Os ilium.
   1. Biventer cervicis.
   2. Complexus cervicis.
   3. Trachelo-mastoideus.
   4. Transversalis cervicis.
   5. Cervicalis ascendens.
   6. Lumbo-costalis.
   7. Longissimus dorsi.
   8. Sacro-lumbalis.
   9. Spinalis dorsi.
   10. Spinalis cervicis.
   11. Semispinalis dorsi.
   12, 12. Levatores costarum.
   13, 13. Intercostales.
   14. Obliquus capitis superior.

Tab.X.

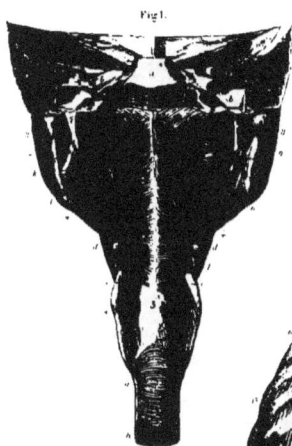

Fig.1.

Fig.2.

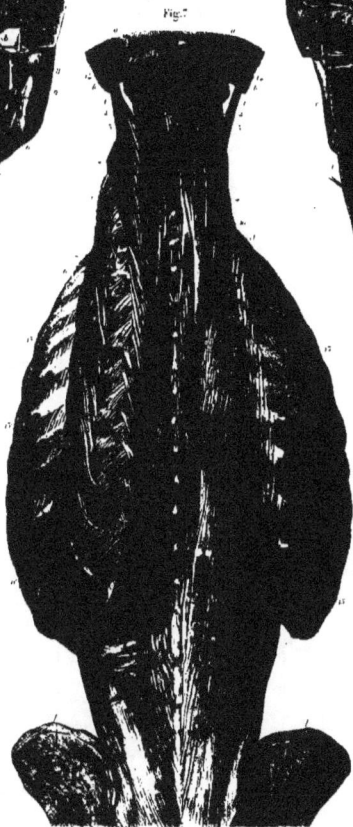

Fig.7.

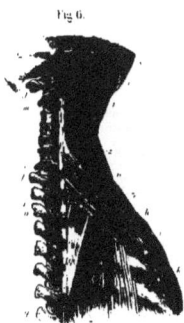

Fig.6.

Fig.3.

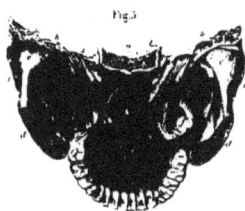

Fig.5.

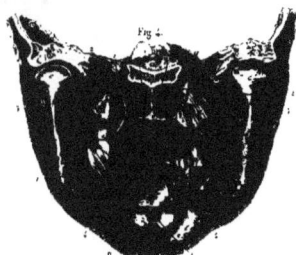

Fig.4.

PLATE XI.

MUSCLES OF THE TRUNK, ARMS, AND FEET.

Fig. 1.

Muscles of Face, Trunk, Arms, and upper part of Thighs—anterior view.

a Occipito-frontalis tendon.
b Malar bone.
c Inferior maxilla.
d Thyroid gland.
e Trachea.
f Clavicle.
g Manubrium of sternum.
h Body of sternum.
i Cormcoid process.
k Acromion.
l First rib (cartilage).
m Second rib (cartilage).
n Third rib (cartilage).
o Fourth rib (cartilage).
p Symphysis pubis.
q Anterior superior spine of ilium.
r Os humeri.
s Interclavicular ligament.
t Rhomboid ligament.
u Ligamenta coruscantia.
v Claviculo-acromial ligament.
w Coraco-acromial ligament.

1. Frontalis.
2. Pyramidalis nasi.
3. Attollens auris.
4. Attrahens auris.
5. Orbicularis palpebrarum.
6. Levator labii superioris alæque nasi with compressor nasi.
7. Levator labii superioris propria.
8. Zygomaticus minor.
9. Zygomaticus major.
10. Levator anguli oris.
11. Masseter.
12. Buccinator.
13. Triangularis menti or depressor anguli oris.
14. Quadratus menti or depressor labii inferioris.
15. Levator menti.
16. Orbicularis oris.
(For other muscles of face, vide Plate IX. Figs. 1 and 4.)
17. Platysma-myoides or latissimus colli.
18. Sterno-cleido-mastoideus.
19. Sterno-hyoideus.
20. Scaleni.
(For muscles of neck, vide Plate IX. Figs. 2, 3, and 4.)
21. Pectoralis major, clavicular and sternal portions.
22. Pectoralis minor.
23. Subclavian.
24. Serratus magnus antica.
25. External oblique (abdominis).
26. Linea alba.
27. Rectus abdominis.
28. Transverse aponeuroses of rectus abdominis.

29. Pyramidalis abdominis.
30. Obliquus internus.
31. Poupart's ligament.
32. External pillar of Poupart's ligament.
33. Internal pillar of Poupart's ligament.
34. External abdominal ring.
35. Internal abdominal ring.
36. Inguinal canal.
37. Deltoid.
38. Coraco-brachialis.
39. Short head of biceps.
40. Long head of biceps.
41. Biceps flexor cubiti muscle and tendinous insertion.
42. Subscapularis.
43. Brachialis.
44. Internal head of triceps.
45. Pronator teres.
46. Supinator longus.
47. Flexor carpi radialis.
48. Palmaris longus.
49. Flexor carpi ulnaris.
50. Flexores digitorum communnes.
51. Flexor pollicis longus.
52. Anterior annular ligament of carpus.
53. Abductor pollicis.
54. Palmaris brevis.
55. Adductor pollicis.
56. Extensor carpi radialis longus.
57. Extensor carpi radialis brevis.
58. Extensor ossis metacarpi pollicis.
59. Extensor primi internodii pollicis.
60. Extensor secundi internodii pollicis.
61. Extensor indicis.
62. Extensor digitorum communis.
63. Abductor indicis.
64. Lumbricales.
65. Abductor of little finger.
66. Fascia lata femoris.
67. External femoral ring.
68. Falciform process of fascia lata.

Fig. 2.

Plantar Fascia or Aponeurosis of right Foot.

a Origin from tuberosity of os calcis.
b Centre or body of fascia.
c Anterior five divisions of fascia.
d Transverse connecting fibres.
e Sheath of abductor and flexor brevis of great toe.
f Sheath of abductor and flexor brevis minimi digiti.
g Deep transverse fibres of fascia.

Fig. 3.

Plantar Muscles, first layer—inferior surface, right Foot.

a Tuberosity of os calcis.
b Tubercle of fifth metatarsal bone.
c Lesser head of first metatarsal bone.

1. Flexor digitorum communis brevis v. perforatus.
2. Tendon of flexor digitorum communis perforans.
3. Abductor pollicis pedis.
4. Flexor brevis pollicis pedis.
5. Abductor minimi digiti.
6. Flexor brevis minimi digiti.
7. Long flexor tendon of great toe.
8. Lumbricales.

Fig. 4.

Second layer of Plantar Muscles of right Foot.

a Tuberosity of os calcis.
b Tubercle of fifth metatarsal bone.
c Tubercle of os naviculare.
d External tarsal ligament.
f Ligaments of lesser heads of metatarsus.

1. Tendon of flexor communis perforans.
2. Musculus accessorius.
3. Lumbricales.
4. Tendon of flexor longus pollicis pedis.
5. Tendon of peroneus longus.
6. Portion of abductor pollicis pedis.
7. Flexor brevis pollicis pedis.
8. Flexor brevis minimi digiti.
9. Interosseus internus tertius.
10. Transversalis pedis.

Fig. 5.

Third layer of Plantar Muscles of right Foot.

a Tuberosity of os calcis.
b Tubercle of fifth metatarsal bone.
c Tubercle of os naviculare
d Internal tarsal ligament.
e Tendon of peroneus longus.

1. Flexor brevis pollicis pedis.
2. Adductor pollicis pedis.
3. Transversalis pedis.
4. Abductor digiti minimi.

Fig. 6.

Fourth layer of Plantar Muscles of right Foot.

a Os calcis.
b Astragalus.
c Tuberosity of fifth metatarsal bone.

1. Extensor brevis pollicis pedis.
2. Extensor digitorum communis brevis.
3. Interosseus externus (1).
4. Interosseus externus (2).
5. Interosseus externus (3).
6. Interosseus externus (4).
7. Abductor digiti minimi.

Tab. XI

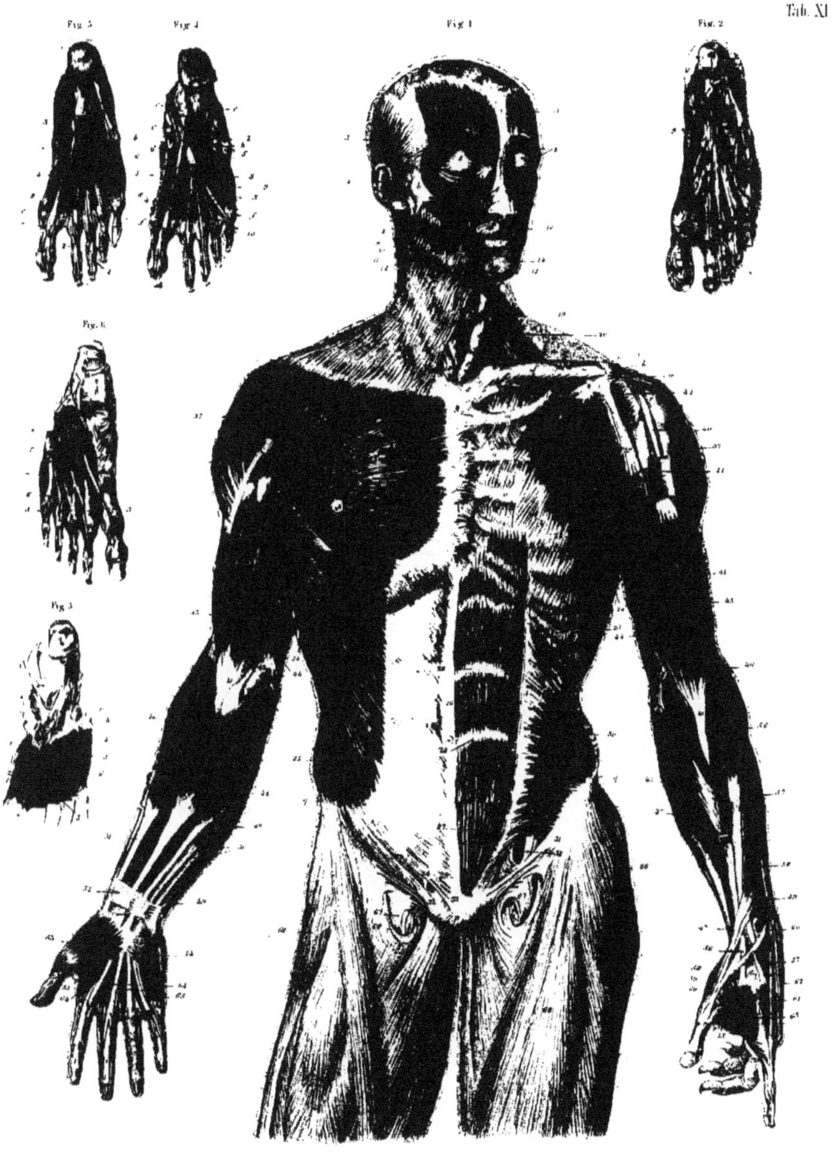

Fig. 5   Fig. 4   Fig 1   Fig. 2

Fig. 6.

Fig. 3

PLATE XII.

MUSCLES OF TRUNK, NECK, AND ARMS.

(Posterior View, with some of Anterior Surface.)

Fig. 1.

Muscles of Trunk, Pelvis, upper part of Thighs, and Arms.

a   Occipito-frontalis tendon or aponeurosis.
b   Superior semicircular line of occiput.
c   Ramus of inferior maxilla.
d   Spinous process of cervical vertebræ.
e   Spinous processes of dorsal vertebræ.
f   Spinous processes of lumbar vertebræ.
g   Os sacrum.
h   Crest of ilium.
i   Tuber ischii.
k   Os ischium.
l   Os ilium.
m   Trochanter major.
n   Clavicle.
o   Spine of scapula.
p   Base of scapula.
q   Acromion.
r   External tuberosity of humerus.
s   Humerus.
t   Olecranon process.
u   Internal condyle of humerus.
v   External condyle of humerus.
w   Ulna.
x   Anterior annular ligament of wrist.
y   Posterior annular ligament of wrist.
z   Last rib.
   1. Frontalis.
   2. Orbicularis palpebrarum.
   3. Attollens auris.
   4. Retrahentes auris.
   5. Attrahens auris.
   6. Masseter.
   7. Occipitales.
   8. Sterno-cleido-mastoideus.
   9. Splenius capitis.
  10. Splenius colli.
  11. Complexus cervicis.
  12. Levator anguli scapulæ.
  13. Trapezius.
  14. Rhomboideus minor.
  15. Rhomboideus major.
  16. Latissimus dorsi.
  17. Serratus posticus inferior.
  18. Serratus anticus major.
  19. Intercostales externi.
  20. Sacro-lumbalis.
  21. Obliquus abdominis externus.
  22. Obliquus abdominis internus.
  23. Glutæus maximus (divided).
  24. Glutæus medius.
  25. Pyriformis.

  26. Gemellus superior.
  27. Obturator internus.
  28. Gemellus inferior.
  29. Quadratus femoris.
  30. Obturator externus.
  31. Vastus externus.
  32. Semimembranosus.
  33. Adductor magnus.
  34. Supraspinatus.
  35. Infraspinatus.
  36. Teres minor.
  37. Teres major.
  38. Deltoideus.
  39. Triceps brachii.
  40. Caput longum tricipitis brachii.
  41. Caput externum tricipitis brachii.
  42. Caput internum tricipitis brachii.
  43. Anconæus.
  44. Brachialis internus.
  45. Supinator longus.
  46. Extensor digitorum communis.
  47. Extensor carpi ulnaris.
  48. Extensores carpi radiales.
  49. Extensor pollicis brevis.
  50. Abductor pollicis longus.
  51. Extensor pollicis longus.
  52. Flexor digitorum communis.

Fig. 2.

Deep Muscles of Neck—anterior view.

a   Body of sphenoid bone.
b   Petrous portion of temporal bone.
c   Anterior tubercle of atlas.
d   Transverse process of atlas.
e   First dorsal vertebra.
f   First rib.
g   Second rib.
   1. Longus colli.
   2. Rectus capitis anticus major.
   3. Rectus capitis anticus minor.
   4. Rectus capitis lateralis.
   5. Scalenus anticus.
   6. Scalenus medius.
   7. Scalenus posticus.
   8. Intertransversarii.

Fig. 3.

Deep Muscles of back of Neck.

a   Occipital bone.
b   Mastoid process.
c   Posterior tubercle of atlas.

d   Transverse process of atlas.
e   Transverse process of dentata.
f   Spinous process of dentata.
g   Spinous process of cervical vertebræ.
   1. Rectus capitis posticus minor.
   2. Rectus capitis posticus major.
   3. Obliquus capitis superior.
   4. Obliquus capitis inferior.
   5. Interspinales.
   6. Multifidus spinæ (cervicis).

Fig. 4.

Tendons and Tendinous Sheaths on posterior surface of Carpus.

a   Ulna.
b   Radius.
c   Posterior common ligament of carpus.
d   Carpus—
   1. Fibrous septum between abductor longus and extensor brevis pollicis.
   2. Fibrous septum between extensor carpi radialis longus and brevis.
   3. Fibrous septum between extensor pollicis longus.
   4. Fibrous septum between extensor digitorum communis and extensor indicis.
   5. Fibrous septum for extensor minimi digiti.
   6. Fibrous septum for extensor carpi ulnaris.

Fig. 5.

Tendons and Tendinous Aponeuroses of right Wrist and Hand—anterior surface.

a   Radius.
b   Os pisiforme.
c   Muscular mass of thumb.
d   Muscular mass of little finger.
   1. Palmaris brevis.
   2. Palmaris longus.
   3. Anterior annular ligament of carpus.
   4. Anterior proper carpal ligament.
   5. Palmar fascia or aponeurosis.
   6. Terminations of palmar aponeurosis.
   7. Transverse palmar ligaments.
   8. Sheaths of long flexor tendons.
   9. Flexor carpi ulnaris tendon.
  10. Flexor carpi radialis tendon.

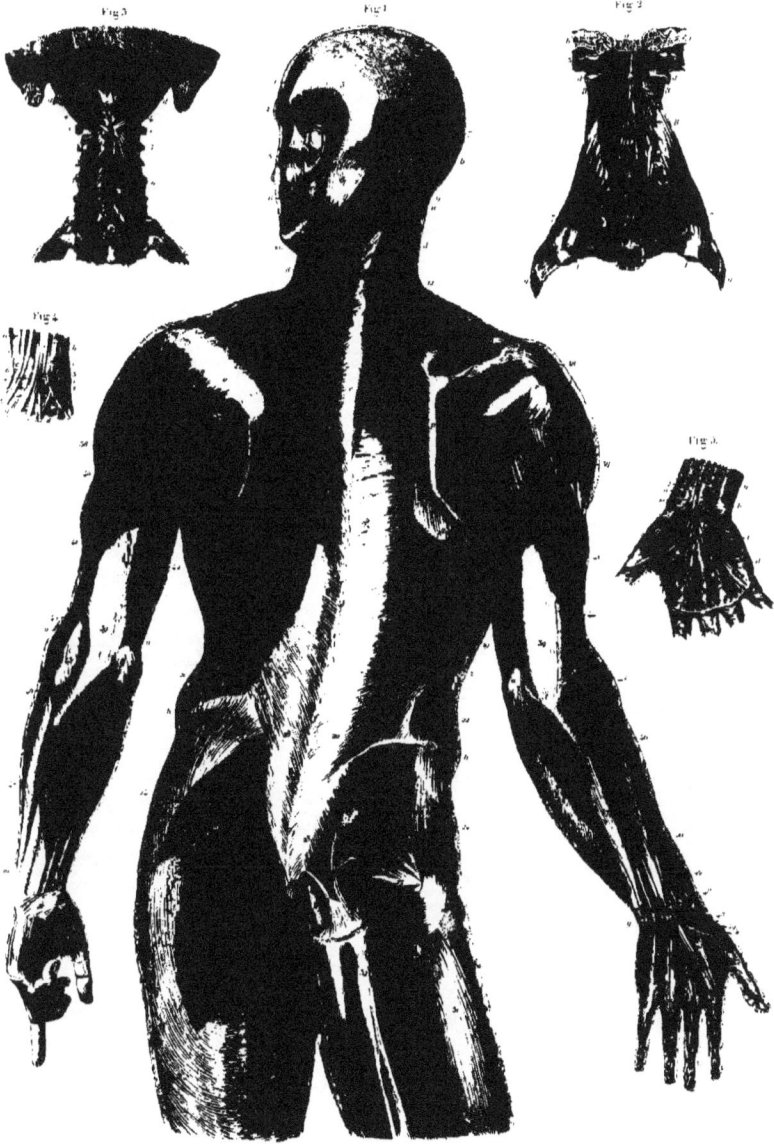

Tab.XII.

Fig.5 Fig.1 Fig.2

Fig.4

Fig.3

PLATE XIII.

MUSCLES OF TRUNK, PERINEUM, FOREARM, AND HAND.

Fig. 1.

Deep Muscles of Abdomen, Diaphragm, and Pelvis.

a Inferior border of thorax.
b Xiphoid process.
c Cut edges of oblique and transversalis muscles.
d Symphysis pubis.
e Horizontal ramus of pubes.
f Lumbar vertebræ.
g Os sacrum.
h Os coccygis.
i Crest of ilium.
  1. Costal or muscular portion of diaphragm.
  2. Tendon of diaphragm.
  3. Internal crus or pillar of diaphragm.
  4. Middle crus or pillar of diaphragm.
  5. External crus or pillar of diaphragm.
  6. Opening for vena cava.
  7. Œsophageal opening.
  8. Aortic opening.
  9. Psoas major.
  10. Psoas minor.
  11. Quadratus lumborum.
  12. Transversalis and fascia transversalis.
  13. Iliacus internus.
  14. Pyriformis.
  15. Levator ani.
  16. Sartorius.
  17. Rectus femoris.
  18. Pectinæus.
  19. Adductor longus.
  20. Tensor fasciæ latæ.
  21. Glutæus medius.

Fig. 2.

Muscles of Ano-Perineal region and upper part of Thigh.

a Tuber ischii.
b Ramus ascendens ischii.
c Great sacro-sciatic ligament.
d Os coccygis.
e Corpus spongiosum urethræ.
f Corpus cavernosum penis.
g Anus.
h Penis.
i Spermatic cord.
  1. Accelerator urinæ or ejaculator seminis.
  2. Erector penis.
  3. Transversus perinæi.
  4. Sphincter ani externus.
  5. Levator ani.

  6. Coccygeus.
  7. Adductor longus.
  8. Gracilis.
  9. Adductor magnus.
  10. Glutæus maximus.

Fig. 3.

Deep Muscles of right Forearm and Hand —anterior or palmar surface.

a Lower end of humerus.
b External condyle of humerus.
c Internal condyle of humerus.
d Capitulum or lesser head of radius.
e Neck of radius.
f Styloid process of radius.
g Styloid process of ulna.
h Os pisiforme.
i Hamular process.
k Interosseous membrane.
  1. Flexor pollicis longus.
  2. Flexor digitorum communis profundus a. perforans.
  3. Pronator quadratus.
  4. Adductor pollicis.
  5. Interosseus internus (2).
  6. Interosseus internus (3).

Fig. 4.

Deep Muscles of right Forearm and Hand —posterior surface.

a Lower end of humerus.
b External condyle of humerus.
c Internal condyle of humerus.
d Olecranon process.
e Lesser head of radius.
f Radius.
g Ulna.
h Carpus.
  1. Brachialis internus.
  2. Anconæus.
  3. Supinator brevis.
  4. Extensor ossis metacarpi pollicis.
  5. Extensor primi internodii pollicis.
  6. Extensor secundi internodii pollicis.
  7. Extensor proprius indicis.
  8. Extensores carpi radiales.
  9. Flexor carpi ulnaris.
  10. Interosseus externus (1).
  11. Interosseus externus (2).
  12. Interosseus externus (3).
  13. Interosseus externus (4).

  14. Adductor pollicis.
  15. Fibrous sheaths for extensor tendons.

Fig. 5.

First or superficial layer of Palmar Muscles of right Hand, beneath Palmar Aponeurosis.

a Os pisiforme.
b Anterior annular ligament of carpus.
c Os metacarpi pollicis.
  1. Abductor pollicis brevis.
  2. Flexor pollicis brevis.
  3. Opponens pollicis.
  4. Adductor pollicis.
  5. Flexor pollicis longus.
  6. Abductor digiti minimi.
  7. Flexor brevis digiti minimi.
  8. Flexores digitorum communes.
  9. Tendons of flexor digitorum sublimis or perforatus.
  10. Tendons of flexor digitorum profundus or perforans.
  11. Lumbricalis (1).
  12. Lumbricalis (2).
  13. Lumbricalis (3).
  14. Lumbricalis (4).
  15. Tendon of flexor carpi radialis.
  16. Pronator quadratus.
  17. Flexor carpi ulnaris.
  18. Interosseus externus (1).

Fig. 6.

Deep Palmar Muscles of left Hand.

a Styloid process of radius.
b Os pisiforme.
c Carpal ligament.
  1. Pronator quadratus.
  2. Opponens pollicis.
  3. Adductor pollicis.
  4. Opponens digiti minimi.
  5. Interosseus externus (1) abductor indicis.
  6. Interosseus internus (1) adductor indicis.
  7. Interosseus internus (3) adductor digiti minimi.
  8. Interosseus externus (4) abductor digiti quarti.
  9. Interosseus internus (3) adductor digiti quarti.
  10. Interosseus externus (3) abductor digiti medii internus.
  11. Interosseus externus (2) abductor digiti medii externus.

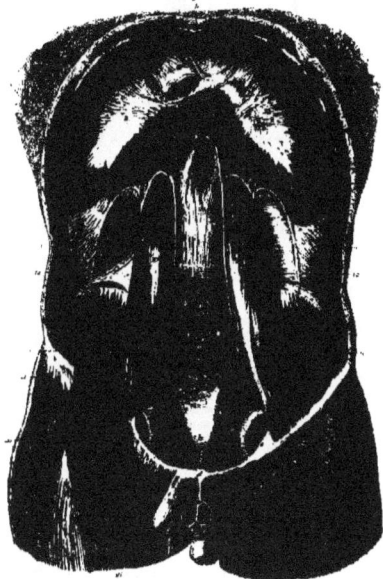

Fig. 1.

Fig. 4.

Fig. 3.

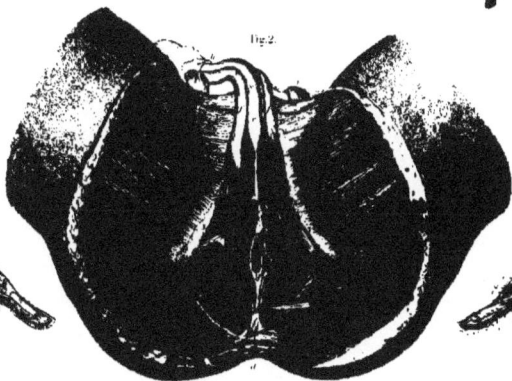

Fig. 2.

Fig. 6.

Fig. 5.

PLATE XIV.

MUSCLES OF THE ANTERIOR AND EXTERNAL SURFACES OF PELVIS AND LOWER EXTREMITIES.

Fig. 1.

Muscles of anterior surface of Lower Extremities.

a   Crest of ilium.
b   Anterior superior spinous process.
c   Trochanter major.
d   Symphysis pubis.
e   Trochanter minor.
f   Patella.
g   Tuberosity of tibia.
h   Tibia.
i   Malleolus internus.
k   Malleolus externus.
l   Anterior annular ligament of ankle-joint.
m   Fibula.
n   Linea alba.
o   Poupart's ligament.
p   Internal pillar of external abdominal ring.
q   External pillar of external abdominal ring.
r   External abdominal ring.
s   Internal abdominal ring.
t   Posterior boundary of inguinal canal.
   1.   Obliquus abdominis externus.
   2.   Transversalis abdominis.
   3.   Tensor fasciæ latæ.
   4.   Glutæus medius.
   5.   Iliacus internus.

   6.   Psoas major.
   7.   Pectinæus.
   8.   Sartorius.
   9.   Adductor longus.
10.   Rectus femoris.
11.   Tendo communis extensorius.
12.   Ligamentum patellæ.
13.   Vastus internus.
14.   Vastus externus.
14*. Tendinous portion of vastus externus.
15.   Gracilis.
16.   Adductor magnus.
17.   Tibialis anticus.
18.   Extensor longus pollicis pedis.
19.   Extensor digitorum communis longus.
20.   Peronæus tertius.
21.   Peronæus longus brevis.
22.   Gastrocnemius.
23.   Soleus.
24.   Extensor brevis pollicis pedis.
25.   Extensor digitorum communis brevis.

Fig. 2.

Muscles on external surface of right side of Pelvis and Lower Extremity.

a   Crest of ilium.
b   Anterior superior spine of ilium.

c   External condyles of knee-joint.
d   Tibia.
e   Patella.
f   Anterior annular ligament of ankle.
g   External portion of annular ligament.
h   Tuberosity of fifth metatarsal bone.
   1.   Tensor fasciæ latæ.
   2.   Fascia lata.
   3.   Glutæus medius.
   4.   Glutæus maximus.
   5.   Sartorius.
   6.   Rectus femoris.
   7.   Vastus externus.
   8.   Biceps femoris (caput longum).
   9.   Caput breve bicipitis femoris.
10.   Tibialis anticus.
11.   Extensor digitorum communis longus.
12.   Extensor longus pollicis pedis.
13.   Peronæus tertius v. parvus.
14.   Peronæus longus v. primus.
15.   Peronæus brevis v. secundus.
16.   Sheaths of long and short peronæal tendons.
17.   Soleus.
18.   Gastrocnemius.
19.   Tendo Achillis.
20.   Extensor digitorum communis brevis.
21.   Abductor digiti minimi.

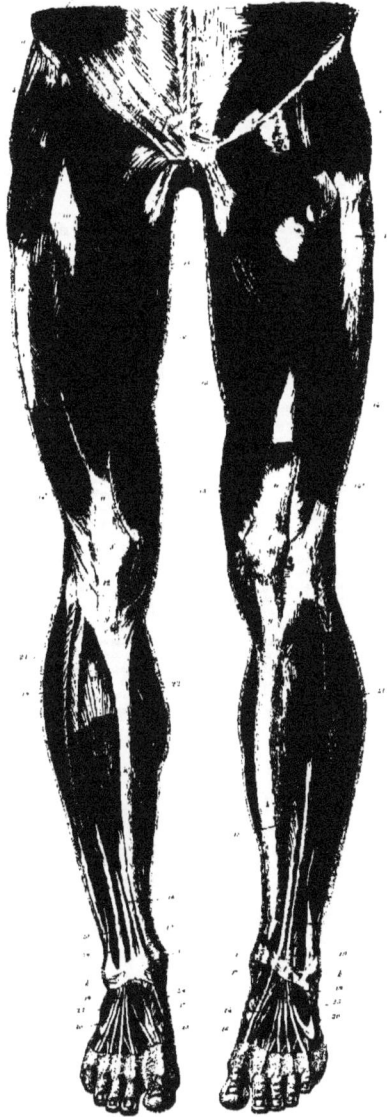

PLATE XV.

MUSCLES OF THE POSTERIOR AND INNER SURFACES OF PELVIS AND LOWER EXTREMITIES.

Fig. 1.

Muscles of posterior surfaces of Pelvis and Lower Extremities.

a Crest of Ilium.
b Os ilium.
o Os coccygis.
d Tuber Ischii.
e Ramus ascendens of ischium.
f Ramus descendens of pubes.
g Trochanter major.
h Os sacrum.
i Lesser sacro-sciatic ligament.
k Greater sacro-sciatic ligament
l Linea aspera.
m Os femoris.
n Popliteal fossa.
o Fibula.
p Malleolus externus.
q Malleolus internus.
r Tendo Achillis.
s Oblique line of tibia.
  1. Glutæus maximus.
  2. Glutæus medius.
  3. Pyriformis.
  4. Gemellus superior.
  5. Obturator internus.
  6. Gemellus inferior.
  7. Quadratus femoris.
  8. Obturator externus.
  9. Caput longum bicipitis femoris.
  10. Caput breve bicipitis femoris.
  11. Tendo bicipitis femoris.
  12. Semitendinosus.
  13. Semimembranosus.

14. Adductor magnus.
15. Openings in adductor magnus for branches of perforating artery and profunda femoris vein.
15.* Inferior opening of Hunter's canal.
16, 16. Gracilis.
17. Sartorius.
18. Vastus externus.
19. Popliteus.
20. Gastrocnemius.
21. Caput externum gastrocnemii.
22. Caput internum gastrocnemii.
23. Plantaris.
24. Tendo plantaris.
25. Tendo Achillis.
26. Soleus.
27. Peronæus longus.
28. Peronæus brevis.
29. Flexor pollicis pedis longus.
30. Tibialis posticus.
31. Flexor communis digitorum pedis longus.

h Lesser sacro-sciatic ligament.
i Great sciatic notch.
k Lesser sciatic notch.
l Ramus descendens pubis.
m Ramus ascendens ischii.
n Anterior sacral foramina.
o Tuber ischii.
p Internal condyles of knee-joint.
q Patella.
r Internal surface of tibia.
s Internal malleolus.
t Internal portion of annular ligament of ankle-joint.
u Glutæus maximus.
  1. Psoas major.
  2. Iliacus internus.
  3. Obturator internus.
  4. Pyriformis.
  5. Sartorius.
  6. Adductor longus v. primus.
  7. Gracilis.
  8. Vastus internus.
  9. Rectus femoris.
  10. Adductor magnus v. tertius.
  11. Semimembranosus.
  12. Semitendinosus.
  13. Gastrocnemius (caput internum).
  14. Soleus.
  15. Tendo Achillis.
  16. Flexor digitorum communis longus per forana.
  17. Flexor pollicis pedis longus.
  18. Tibialis posticus.
  19. Tendo tibialis antici.
  20. Tendo extensoris pollicis pedis longi.
  21. Adductor pollicis pedis.

Fig. 2.

Muscles of inner surface of Pelvis, Thigh, Leg, and Foot.

a Crest of ilium.
b Os sacrum.
c Os coccygis.
d Linea innominata interna.
e Symphysis pubis.
f Foramen obturatorium.
g Great sacto-sciatic ligament.

Tab XV

Fig 1.

Fig 2.

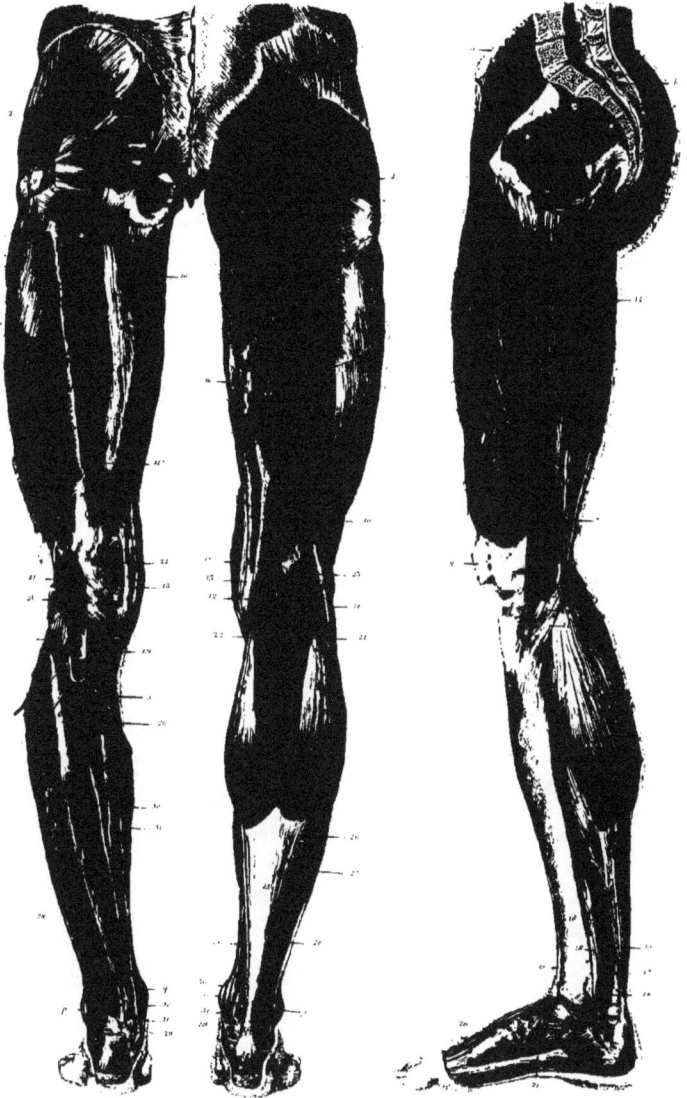

PLATE XVI.

THE HEART, ITS CAVITIES AND VALVES.

Fig. 1.

Anterior surface of Heart and Pericardial covering.

a   Right auricular appendix.
b   Left auricular appendix.
c   Right ventricle.
d   Left ventricle.
e   Transverse or auriculo-ventricular groove.
f   Anterior longitudinal sulcus.
g   Apex cordis.
h   Pericardium divided and thrown back.
   1. Arteria pulmonaria.
   2. Aorta ascendens.
   3. Art. coronaria cordis dextra.
   4. Ramus anterior arteriæ coronariæ cordis sinistræ.
   5. Commencement of great coronary vein of heart.

Fig. 2.

Posterior surface of Heart, Auricles, and Ventricles.

a   Left auricle.
b   Left pulmonary veins.
c   Right pulmonary veins.
d   Right auricle.
e   Opening of inferior vena cava.
f   Left ventricle.
g   Right ventricle.
h   Transverse groove.
i   Posterior longitudinal sulcus.
k   Apex cordis.
l   Edge of divided pericardium.
   1. Art. coronaria cordis dextra.
   2. Ramus descendens art. coronariæ dextræ.
   3. Ramus posterior art. coronariæ sinistræ.
   4. Vena coronaria magna cordis.
   5. Vena coronaria cordis media.
   6. Vena coronaria cordis dextra.

Fig. 3.

Internal cavities of Ventricles—anterior view.

a   Right auricle.
b   Right auricular appendix.
c   Vena cava superior.
d   Vena cava inferior.
e   Left auricle.
f   Left auricular appendix.
g   Venæ pulmonares.

h   Arteria pulmonaria.
i   Aorta ascendens.
k   Right ventricle.
l   Left ventricle.
m   Apex cordis.
n   Septum ventriculorum.
o   Opening of pulmonary artery.
p   Opening of aorta.
q   Tricuspid or right auriculo-ventricular valve.
r   Bicuspid or left auriculo-ventricular valve.
s   Chordæ tendineæ.
t   Musculi pectinati.
u   Fleshy surface of cut edge of right ventricle.

Fig. 4.

Anterior surface of Heart—interior of right auricle exposed.

a   Inter-auricular septum.
b   Vena cava superior.
c   Vena cava inferior.
d   Eustachian valve.
e   Fossa ovalis.
f   Opening of great coronary vein.
g   Right auricular appendix and musculi pectinati.
h   Right auriculo-ventricular opening.
i   Right ventricle.
k   Left ventricle.

Fig. 5.

Interior of right Auricle, Ventricle, and Pulmonary Artery.

a   Right auriculo-ventricular opening.
b   Anterior border of tricuspid valve.
c   Inner border of tricuspid valve.
d   Posterior border of tricuspid valve.
e   Chordæ tendineæ.
f   Musculi pectinati.
g   Carneæ columnæ.
h   Pulmonary artery.
i   Anterior semilunar valve of pulmonary artery.
k   Right semilunar valve of pulmonary artery.
l   Left semilunar valve of pulmonary artery.
m   Corpora aurantii.
n   Sinuses Valsalvæ.
o   Septum ventriculorum.
p   Left ventricle.
q   Ascending aorta.

Fig. 6.

Exterior of left Ventricle and of Aorta.

a   Left auriculo-ventricular opening.
b   Right anterior border of mitral or bicuspid valve.
c   Left posterior border of mitral or bicuspid valve.
d   Origins of chordæ tendineæ.
e   From carneæ columnæ.
f   Septum ventriculorum.
g   Origin of aorta.
h   Posterior semilunar valve of aorta.
i   Right semilunar valve of aorta.
k   Left semilunar valve of aorta.
l   Openings of coronary artery of heart.
m   Sinuses Valsalvæ.
n   Corpora aurantii.

Fig. 7.

Transverse section of Auricles, Aorta, and Pulmonary Artery, immediately above the origins of these vessels, showing the auriculo-ventricular and arterial valves in action.

a   Septum auriculorum.
b   Wall of left auricle.
c   Wall of right auricle.
d   Tricuspid valve.
e   Bicuspid valve.
f   Semilunar valves of pulmonary artery.
g   Semilunar valves of aorta.
h   Left auricular appendix.
i   Right auricular appendix.
   1. Right coronary artery.
   2. Left coronary artery.

Fig. 8.

Transverse section of Ventricles.

   1. Right Ventricle.
   2. Left Ventricle.
   3. Septum ventriculorum.
   4. Cut edge of left ventricle.
   5. Cut edge of right ventricle.
   6. Adipose tissue.
   7. Descending branch of left coronary artery in anterior longitudinal sulcus, with its accompanying vein.
   8. Right coronary artery in posterior longitudinal sulcus with accompanying vein.

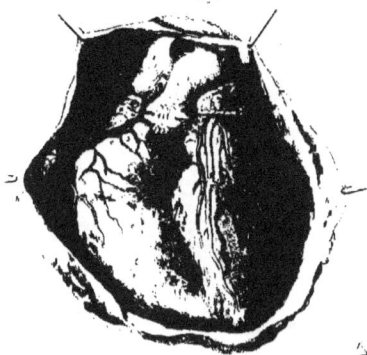

Fig 1

Fig 2

Fig 7

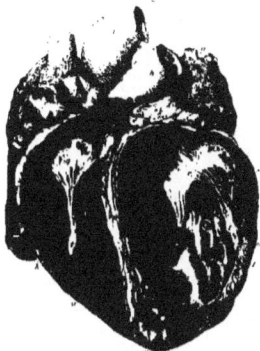

Fig 5

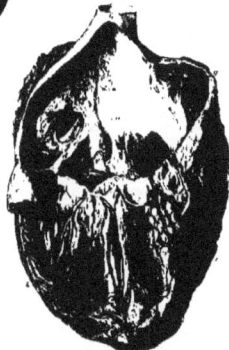

Fig 4

Fig 8

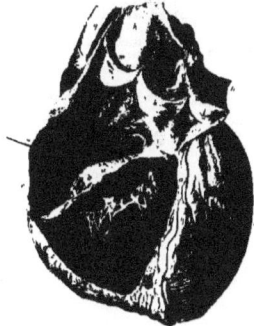

Fig 3

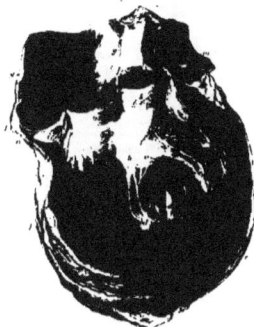

Fig 6

PLATE XVII.

BLOOD-VESSELS (ARTERIES AND VEINS) OF HEAD AND NECK.

Fig. 1.

Arteries of anterior surface of Head and Neck.

a Muscl. occipito-frontalis.
b Mscl. orbicularis palpebrarum.
c M. corrugator supercilii.
d M. levator labii superioris alaeque nasi.
e M. levator labii superioris proprius.
f M. zygomaticus minor.
g M. zygomaticus major.
h M. masseter.
i M. buccinator.
k M. orbicularis oris.
l M. triangularis menti.
m M. quadratus menti.
n M. levator anguli oris.
o M. sterno-cleido-mastoideus.
p M. sterno-hyoideus.
q Glandula thyroidea.
r Trachea.
s Larynx.
t M. cucullaris v. trapezius.
u M. omo-hyoideus.
v M. scalenus anticus.
w M. scalenus medius.
x Clavicula.
   1. Arteria subclavia.
   2. Art. mammaria interna.
   3. Art. transversa scapulae.
   4. Art. transversa colli.
   5. Art. cervicalis ascendens.
   6. Art. thyroidea inferior.
   7. Art. carotis communis.
   8. Art. thyroidea superior.
   9. Art. maxillaris externa v. labialis.
  10. Art. coronaria labii inferioris.
  11. Art. coronaria labii superioris.
  12. Art. angularis.
  13. Artt. dorsales nasi.
  14. Artt. alares nasi.
  15. Art. ophthalmica (with artt. palpebrales).
  16. Art. frontalis.
  17. Art. supraorbitalis.
  18. Art. infraorbitalis.
  19. Artt. temporales profundae (from the art. maxillaris interna).
  20. Art. temporalis (superficialis).
  21. Ramus frontalis art. temporalis.

Fig. 2.

Arteries and Veins of lateral surface of Head, Face, and Neck.

a Musculus platysma-myoides v. latissimus colli.
b Mscl. cucullaris v. trapezius.
c M. deltoideus.
d M. sterno-cleido-mastoideus.
e M. splenius capitis.
f M. splenius colli.

g M. occipitalis.
h Mm. retrahentes auris.
i M. attollens auris.
k M. masseter.
l M. buccinator.
m M. zygomaticus major.
n M. zygomaticus minor.
o M. orbicularis oris.
p M. triangularis menti.
q M. quadratus menti.
r M. orbicularis palpebrarum.
s M. frontalis.
t M. levator labii superioris alaeque nasi.
u Maxilla inferior.
v M. digastricus maxillae inferioris.
w M. milo-hyoideus.
x M. sterno-hyoideus.
y M. omo-hyoideus.
   1. Vena jugularis externa.
   2. Ven. occipitalis.
   3. Ramus communicans inter ven. jugularis externa et interna.
   4. Ven. jugularis interna.
   5. Ven. facialis anterior.
   6. Ven. labialis.
   7. Ven. angularis.
   8. Ven. temporalis.
   9. Ven. ophthalmica cerebralis.
  10. Ven. frontalis.
  11. Arteria carotis externa.
  12. Art. auricularis posterior.
  13. Art. temporalis (superficialis).
  14. Art. transversa faciei.
  15. Art. maxillaris externa, labialis v. facialis.
  16. Art. submentalis.
  17. Art. angularis.
  18. Art. frontalis.

Fig. 3.

Arteries of right side of Neck.

a Maxilla inferior.
b Os hyoides.
c Clavicula.
d Larynx.
e Glandula thyroidea.
f Trachea.
g Acromion scapulae.
h Processus mastoideus.
i Processus styloideus.
k Processus transversus atlantis.
l Musculus digastricus (venter anterior.)
m Mscl. mylo-hyoideus.
n M. hyo-glossus.
o M. stylo-glossus.
p M. sterno-cleido-mastoideus.
q M. levator anguli scapulae.
r M. scalenus anticus.
s M. scalenus medius.
t M. omo-hyoideus.
u M. sterno-hyoideus.
v M. thyro-hyoideus.

w Pharynx.
x Oesophagus.
y Mscl. subclavius.
z M. pectoralis major.
   1. Art. carotis communis dextra.
   2. Bifurcatio art. carotis communis dextrae.
   3. Art. carotis externa.
   4. Art. carotis interna.
   5. Art. thyroidea superior.
   6. Art. laryngea superior.
   7. Art. lingualis.
   8. Ramus hyoideus art. lingualis.
   9. Art. maxillaris externa v. facialis.
  10. Art. palatina ascendens.
  11. Art. submentalis.
  12. Art. occipitalis (with ascending and descending branches).
  13. Art. auris posterior.
  14. Art. temporalis (superficialis).
  15. Art. subclavia dextra.
  16. Truncus thyro-cervicalis.
  17. Art. thyroidea inferior.
  18. Art. cervicalis ascendens.
  19. Art. transversalis humeri.
  20. Art. transversalis colli.
  21. Art. axillaris.
  22. Artt. thoracicae externae.

Fig. 4.

Arteries and Veins of right side of Neck.

a—x as in Fig. 3.
y First rib
  First bone of sternum.
   1. Vena cava superior.
   2. Ven. innominata sinistra.
   3. Ven. innominata dextra.
   4. Ven. subclavia dextra.
   5. Ven. axillaris.
   6. Ven. jugularis externa.
   7. Ven. jugularis interna.
   8. Ven. facialis.
   9. Ven. maxillaris interna.
  10. Ven. jugularis media.
  11. Arcus aortae.
  12. Arteria innominata.
  13. Carotis communis dextra.
  14. Art. subclavia dextra.
  15. Art. axillaris.
  16. Art. carotis externa.
  17. Art. carotis interna.
  18. Art. thyroidea superior.
  19. Art. lingualis.
  20. Art. maxillaris externa v. facialis.
  21. Art. temporalis.
  22. Art. auricularis posterior.
  23. Art. occipitalis.
  24. Art. thyroidea inferior.
  25. Art. transversalis humeri.
  26. Art. transversalis colli.
  27. Artt. thoracicae externae.

Tab. XVII.

Fig. 1.

Fig. 2.

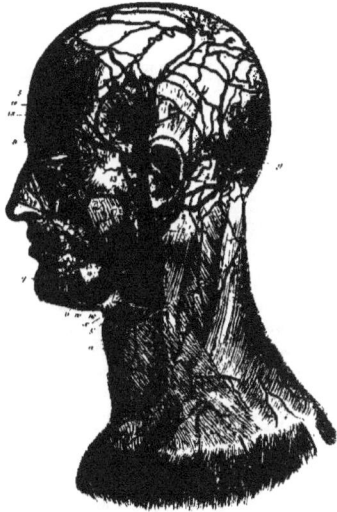

Fig. 3.

Fig. 4.

A. Krause in Leipzig.

PLATE XVII.*

ARTERIES OF HEAD, NECK, AND PELVIS.

Fig. 1.

Internal Maxillary Artery and branches.

a Ramus of inferior maxilla divided, and superior half removed.
b Zygomatic arch divided and malar part removed.
c Tuber maxillare.
d Masseter divided.
e Temporalis divided.
f Pterygoideus externus divided.
g Pterygoideus internus.
h Buccinator.
i Levator labii superioris proprius.
k Zygomaticus minor.
l Zygomaticus major.
m Levator labii superioris alæque nasi.
n Levator anguli oris.
o Compressor nasi.
p Depressor labii inferioris (v. quadratus menti.)
q Depressor labii inferioris (v. quadratus menti.)
r Orbicularis oris.
s Sphincter palpebrarum.
t Ductus Stenonianus divided.
u Sinus longitudinalis superior.
  1. Arteria maxillaris externa.
  2. Ramus muscularis.
  3. Art. coronaria labii inferioris.
  4. A. coronaria labii superioris.
  5. A. angularis.
  6. A. palpebralis superior.
  7. A. infraorbitalis.
  8. A. carotis externa.
  9. A. carotis interna.
  10. Ramus muscularis.
  11. A. auris posterior.
  12. A. auricularis profunda.
  13. A. temporalis superficialis.
  14. A. maxillaris interna.
15. 15. A. meningea media.
16. 16. Artt. temporales profundæ (anterior et posterior).
  17. A. alveolaris inferior.
  18. Ramus buccinatorius.
  19. A. alveolaris superior posterior.
  20. A. infraorbitalis.
  21. A. sphenopalatina.
  22. A. pterygopalatina.

Fig. 2.

The right Subclavian Artery and Vertebral branch.

a First rib.
b M. sterno-cleido mastoideus.
c Scalenus anticus.
d Scalenus medius.
e Longus colli.
f Semispinalis cervicis.
g Obliquus capitis inferior.
h Rectus capitis posticus major.
i Obliquus capitis superior.
k Processus transversus of Atlas.
l Processus transversus 6. vertebra.
  1. Arteria innominata.
  2. A. carotis communis dextra.
  3. A. subclavia dextra.
  4. A. mammaria interna.
  5. A. axis thyroidea.
  6. A. thyroidea inferior.
  7. Divided trunks of cervicalis ascendens transversalis humeri and superficial cervical arteries.
8. 8. A. vertebralis.
  9. Truncus costo-cervicalis.
  10. A. intercostalis suprema.
  11. A. cervicalis profunda.
  12. A. transversalis colli.
  13. A. occipitalis.

Fig. 3.

Right Subclavian Artery viewed from interior of Thorax.

a Sixth cervical vertebra.
b First dorsal vertebra.
c Fourth dorsal vertebra.
d First rib.
e Third rib.
f Clavicula.
g Sternum.
h Mm. intercostales interni.
  1. Arteria subclavia dextra.
  2. A. vertebralis, entering foramen in transverse process of sixth cervical vertebra.
  3. Truncus costo-cervicalis.
  4. A. mammaria interior.
  5. A. intercostalis superior.
  6. Ramus dorsalis of first and second intercostal arteries.
  7. A. intercostalis prima.
  8. A. intercostalis secunda.
  9. Superior and inferior branches of A. intercostalis secunda.
  10. A. intercostalis tertia.
  11. A. intercostalis quarta.
  12. Rami dorsales of third and fourth intercostals.
  13. A. mammaria interna.
  14. Branch of A. mammaria interna which winds upwards above the clavicle.
  15. Rami sternales.
  16. Arteriæ intercostales anteriores.

Fig. 4.

Ophthalmic Artery with its superior branches.

a. a Nervus opticus.
b M. levator palpebrae superioris.
c M. rectus externus.
d M. orbicularis palpebrarum.
e M. obliquus superior.
f Trochlea for tendon of m. obliquus superior.
g Tendon of m. obliquus superior.
i Glandula lacrymalis.
  1. Arteria carotis interna.
2. 2. A. ophthalmica.
  3. A. lacrymalis.
  4. Ramus muscularis of m. rectus externus.
  5. Ramus palpebralis.
  6. Ramus muscularis for m. obliquus superior.
  7. Ramus muscularis for m. levator palpebrae superioris.
  8. A. ethmoidalis posterior.
  9. A. ethmoidalis anterior.
  10. A. supraorbitalis.
  11. A. dorsalis nasi.
  12. A. frontalis.
  13. A. palpebrae superioris.

Fig. 5.

Ophthalmic Artery with its inferior branches.

a. a Nervus opticus.
b. b Musculus rectus superior.
c M. rectus externus.
d Cornea.

e M. obliquus superior.
f Trochlea.
g Tendon of obliquus superior.
h Bulbus oculi.
i M. rectus inferior.
  1. Arteria carotis interna.
  2. Art. ophthalmica.
  3. A. centralis (divided).
  4. A. centralis retinæ.
  5. Divided branch to rectus superior.
  6. Branch to rectus superior.
7. 7. Artt. ciliares posticæ.
  8. A. ethmoidalis posterior.
  9. Branch to superior oblique m. and anterior ethmoidal artery.
  10. Arteria supraorbitalis.
  11. Ophthalmic artery divided anteriorly.

Fig. 6.

Internal view of Pelvic Arteries, viewed from above, with Epigastric and Spermatic branches.

a Intervertebral surface of fourth lumbar vertebra.
b Crest of ilium.
c Horizontal ramus of pubes.
d Psoas muscle.
e Iliacus internus muscle.
f Rectum.
g Vesica urinaria.
h Vas deferens.
i Gimbernaut's ligament.
k Superior opening of femoral ring.
l Portion of fascia iliaca.
m Internal abdominal ring.
n External oblique muscle.
o Internal oblique muscle.
p Transversalis abdominal muscle.
q Fascia transversalis.
r Sheath of rectus muscle.
s Rectus muscle.
t Linea alba.
u Semicircular margin of fascia transversalis bounding internal abdominal ring.
v Tendon of rectus abdominis muscle.
w Tendinous attachment of linea alba.
x Nervus genito-cruralis.
y Nervus spermaticus externus.
z Nervus lumbo-inguinalis.
  1. Aorta abdominalis.
  2. Art. iliaca communis sinistra.
  3. A. iliaca communis dextra.
  4. A. sacra media.
  5. A. iliaca interna v. hypogastrica.
  6. A. iliaca externa.
  7. A. circumflexa ilii.
  8. A. epigastrica inferior.
  9. Ramus anastomoticus pubicus.
  10. A. spermatica interna.

Fig. 7.

Arteries and Veins of Pelvic Cavity in the male subject.

a Fourth lumbar vertebra.
b Promontory of sacrum.
c Os coccygis.
d Internal semicircular ridge.
e Vesica urinaria.
f Vasa deferentia.
g Vesiculæ seminales.
h Right ureter.
i Internal abdominal ring.
k Rectus abdominis muscle.

l Crescentic edge of fascia transversalis.
m Fascia transversalis.
n Fascia iliaca.
o Obturator canal.
p Pelvic fascia of m. obturator internus.
q Arch of pelvic fascia.
r Pelvic fascia of mm. levator ani and coccygeus.
s Pyriform muscle.
t Sacral plexus.
u Rectum.
v Gimbernaut's ligament.
  1. Arteria iliaca communis dextra.
  2. Art. iliaca externa.
  3. Vena iliaca externa.
  4. Art. epigastrica inferior, with vena comites.
  5. A. spermatica interna.
  6. A. iliaca interna s. hypogastrica.
  7. Vena hypogastrica.
  8. A. glutæa superior.
  9. A. vesicalis superior.
  10. A. umbilicalis.
  11. A. obturatoria.
  12. A. vesicalis inferior.
  13. A. hæmorrhoidalis media.
  14. A. hæmorrhoidalis interna.
  15. A. pudenda communis v. interna.
  16. A. glutæa inferior.
  17. Artt. sacrales laterales (superior and inferior).

Fig. 8.

Arteries of Pelvic Cavity in the female subject.

a Last lumbar vertebra.
b Promontory of sacrum.
c Os coccygis.
d Internal semicircular ridge.
e Vesica urinaria.
f Ligamentum teres uteri.
g Uterus.
h Fallopian tube.
i Corpus fimbriatum.
k Ovarium.
l Rectum.
m Musculus coccygeus.
n M. obturator internus.
o Symphysis ossium pubis.
p Musc. pyriformis.
q M. iliacus internus.
r M. iliacus internus.
s Musc. psoas major.
t M. transversus abdominis.
  1. Aorta abdominalis.
  2. Arteria mesenterica inferior.
  3. Art. lumbalis (3).
  4. A. iliaca communis sinistra.
  5. A. iliaca communis dextra.
  6. A. sacralis media.
  7. A. iliaca externa.
  8. A. circumflexa ilii.
  9. A. epigastrica inferior.
  10. Ramus anastomoticus pubicus.
  11. A. hypogastrica.
  12. A. ileo-lumbalis.
  13. A. Obturatoria.
  14. A. umbilicalis (obliterated).
  15. A. vesicalis.
  16. A. glutæa superior.
  17. A. sacralis lateralis superior and inferior.
  18. A. uterina.
  19. A. pudenda communis.
  20. A. hæmorrhoidalis media.
  21. A. ischiadica v. glutæa inferior.

Fig. 2.

Fig. 3.

Fig. 1.

Fig. 4.

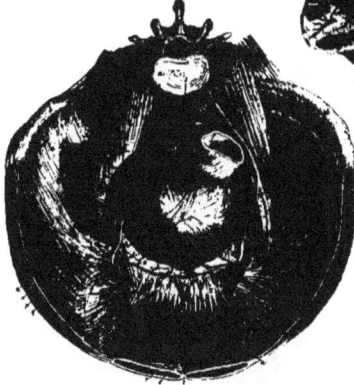

Fig. 6.

Fig. 5.

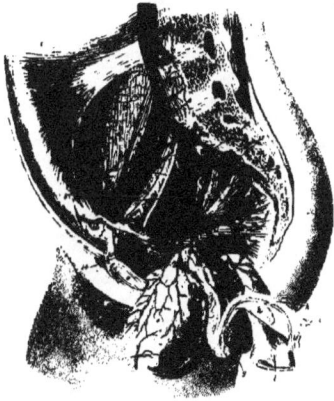

Fig. 8.

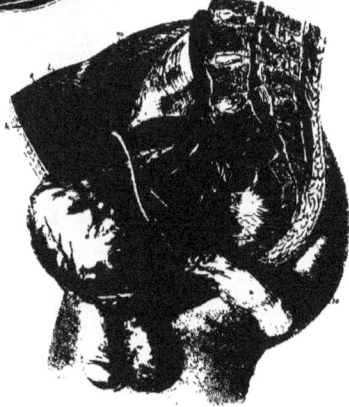

Fig. 7.

PLATE XVIII.

ARTERIES OF ANTERIOR SURFACE OF TRUNK, ARM, FOREARM, AND HAND.

Fig. 1.

Arteries of Trunk, Axilla, and Inguinal Regions.

a Clavicle.
b First rib.
c Last rib.
d Sternum.
e Xiphoid process.
f Deltoid muscle.
g Insertion of pectoralis major muscle.
g' Clavicular origin of pectoralis major muscle.
g" Sternal origin of pectoralis major muscle.
h Biceps flexor cubiti muscle.
i Coraco-brachialis muscle.
k M. pectoralis minor.
l M. subscapularis.
m M. latissimus dorsi.
n M. scalenus anticus.
o M. serratus anticus major.
p Min. intercostales interni.
p' M. intercostalis externus primus.
q M. triangularis sterni.
r M. sternohyoideus.
s M. sternal and
s' Clavicular insertion of m. sterno-cleido-mastoideus.
t M. omohyoideus.
u M. trapezius.
v M. subclavius.
w Pleura costalis.
x M. obliquus abdominis externus.
y M. obliquus abdominis internus.
z M. transversalis abdominis.
z M. rectus abdominis.
β Sheath of rectus muscle cut through and reflected.
γ Umbilicus.
δ Linea alba.
ε Cut edge of obliquus internus muscle.
ζ Cut edge of rectus sheath.
η Posterior layer of rectus sheath.
λ Fascia transversalis.
λ Poupart's ligament.
μ External abdominal ring.
ν Spermatic cord.
π Suspensory ligament of penis.
ρ Fascia lata.
σ Falciform process of fascia lata.
τ Fossa ovalis or femoral ring of fascia lata.
φ Sartorius muscle.
φ Psoas muscle.
ω Pectineus muscle.
1. Arteria innominata.
2. Art. carotis communis dextra.
3. Art. subclavia dextra.
4. Axis thyroidea.
5. Art. mammaria interna.
6. Art. axillaris.
7. Art. thoracico-acromialis, divided.
8. Art. thoracica longa.
9. Art. circumflexa posterior.
10. Art. subscapularis.
11. Art. circumflexa scapula.
12. Art. thoracica dorsalis of the art. subscapularis.
13. Artt. intercostales anteriores.
14. Art. epigastrica superior.
15. Rami perforantes of art. mammaria interna.
16. Art. thoracica secunda v. thoracico-acromialis.
17. Art. thoracica prima.
18. Muscular branch of thoracico-acromialis art.
19. Ramus acromialis of thoracico-acromialis art.
20. Ramus deltoideus of thoracico-acromialis art.
21. Rami musculares of artt. epigastricon.
22. Art. epigastrica inferior.
23. Art. circumflexa ilii.
24. Art. cruralis v. femoralis.
25. Art. epigastrica superficialis (divided).
26. Art. spermatica interna.
27. Junction of right and left vena innominata to form vena cava superior.
28. Vena innominata dextra.
29. Vena jugularis interna.
30. Vena subclavia.
31. Vena axillaris.

32. Vena cephalica.
33. Vena cruralis v. femoralis.
34. Vena saphena magna.
35. Vena dorsalis penis.
36. Plexus brachialis.
37. Nervus medianus, its double origin perforated by axillary artery.
38. Nerv. suprascapularis.
39. Nerv. musculo-cutaneus v. perforans Casseri.
40. Nerv. axillaris.
41. Nerv. cutaneus internus (major).
42. Nerv. ulnaris.
43. Nerv. radialis.
44. Pars infraclavicularis plexus brachialis.

Fig. 2.

Superficial Arteries on the internal and anterior surface of Arm, Forearm, and Hand.

a Musculus deltoideus.
b M. pectoralis major.
c M. latissimus dorsi.
d M. biceps flexor cubiti.
e Semilunar fascia of biceps.
f M. coraco brachialis.
g Long head of triceps extensor cubiti m.
h Short head of triceps extensor cubiti m.
i Brachialis anticus m.
k Internal intermuscular ligament.
j Internal condyle of humerus.
m M. supinator longus.
n M. pronator teres.
o M. flexor carpi ulnaris.
p M. palmaris longus.
q M. flexor carpi ulnaris.
r M. extensor carpi radialis longus.
s M. flexor pollicis longus.
t M. flexor digitorum communis sublimis.
t* M. flexor digitorum communis profundus.
u M. abductor pollicis longus.
v M. extensor pollicis brevis.
w Anterior annular ligament of wrist.
x Ball of thumb, abductor and flexor brevis pollicis.
y Tendon of flexor longus pollicis.
z M. adductor pollicis.
α Tendones of flexor digitorum communis.
β M. palmaris brevis.
γ Mm. abductor and flexor brevis minimi digiti.
δ Mm. lumbricales.
ε Ligamentary sheaths of tendons.
ζ Tendon of flexor digitorum communis sublimis to little finger.
η Tendon of flexor digitorum communis profundus.
1. Arteria brachialis.
2. Rami musculares to m. coraco-brachialis and biceps brachii.
3. Rami musculares to m. triceps brachii.
4. Art. profunda superior brachii.
5. Art. anastomotica magna.
6. Art. ulnaris.
7. Art. radialis.
8. Art. recurrens radialis.
9. Ramus dorsalis arteriæ radialis.
10. Ramus volaris art. radialis.
11. Ramus muscularis to ball of thumb.
12. Art. pollicis radialis from art. princeps pollicis.
13. Art. pollicis ulnaris from art. princeps pollicis.
14. Art. indicis radialis from art. princeps pollicis.
15. Ramus volaris arteriæ ulnaris.
16. Arcus palmaris superior.
17. Artt. digitales volares communes.
18. Art. ulnaris volaris.
19. Art. radialis dorsalis.
20. Ramus profundus v. communicans.

Fig. 3.

Deep Arteries of Arm, Forearm, and Hand—anterior surface.

a Musculus coraco-brachialis.
b M. latissimus dorsi.

c Caput longum musculi tricipitis.
d Caput breve musculi tricipitis.
e M. brachialis anticus.
f M. supinator brevis.
g Ligamentum intermusculare internum.
h Condylus internus humeri.
i Tendon of biceps (divided).
k M. extensor carpi radialis longus.
l M. extensor carpi radialis brevis.
m Tendon of supinator longus (divided).
n Radial insertion of pronator teres.
o Origin of mm. radialis internus and palmaris longus.
p Interosseous membrane.
q Flexor longus pollicis.
r Flexor muscle (divided).
s M. pronator quadratus.
t Tendon of flexor carpi ulnaris (divided).
u Anterior annular ligament (divided).
w Opponens digiti minimi.
x Mm. interossei.
1. Art. brachialis.
2. Art. profunda superior brachii.
3. Art. anastomotica magna.
4. Art. recurrens radialis.
5. Art. radialis.
6. Art. ulnaris.
7. Art. ulnaris recurrens anterior.
8. Art. ulnaris recurrens posterior.
9. Art. ulnaris recurrens posterior.
10. Arteria interossea.
11. Continuation of ulnar artery.
12. Ramus dorsalis arteriæ radialis.
13. Superficialis volæ.
14. Ramus dorsalis arteriæ ulnaris.
15. Section of communicating branch of ulnar artery.
16. Deep palmar arch.
17. Ramus profundus of ulnar artery.
18. Art. princeps pollicis.
19. Art. indicis radialis.
20. Art. digitales communes (divided).
21. Artt. interosseæ palmares.

Fig. 4.

Arteries of posterior surface of lower part of Arm, Forearm, and Hand.

a Insertion of deltoid muscle.
b Long head of triceps muscle.
c Short head of triceps muscle.
d Olecranon.
e Ulna.
f Musc. biceps flexor cubiti.
g M. brachialis internus.
h M. supinator longus.
i M. extensor carpi radialis longus.
k M. extensor carpi radialis brevis.
l M. extensor digitorum communis.
m External condyle of humerus.
n M. anconæus.
o M. extensor carpi ulnaris.
p M. flexor carpi ulnaris.
q M. extensor ossis metacarpi pollicis.
r M. extensor primi internodii pollicis.
s M. extensor secundi internodii pollicis.
t Posterior annular ligament of wrist.
u Exterior tendon of ring finger.
v M. interosseus externus primus.
1. Rami musculares of the art. profunda brachii.
2. Art. superior profunda brachii.
3. Branch of radial recurrent artery.
4. Art. recurrens interossea.
5. Rete articulare cubiti.
6. Branch of the art. interossea externa.
7. Branch of the art. interossea interna.
8. Art. interossea externa inferior.
9. Ramus dorsalis arteriæ ulnaris.
10. Ramus dorsalis arteriæ ulnaris.
11. Branch of dorsal carpal artery.
12. Art. dorsalis pollicis ulnaris.
13. Art. dorsalis indicis radialis.
14. Art. pollicis ulnaris.
15. Art. dorsalis digiti (3) ulnaris.
16. Art. interossea metacarpi (3) et (1).
17. Art. digiti (5) radialis dorsalis.
18. Art. digiti (4) ulnaris dorsalis.

Fig. 1.

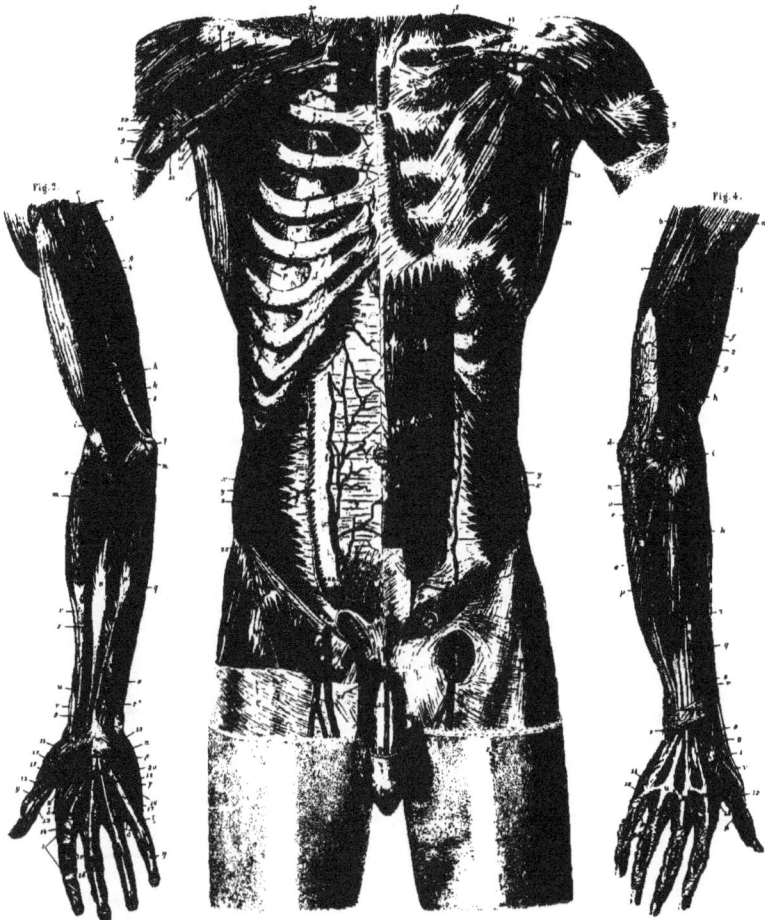

Fig. 2.

Fig. 4.

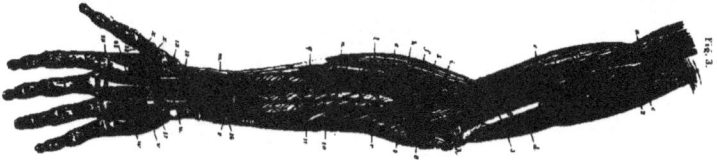

Fig. 3.

PLATE XIX.

BLOOD-VESSELS OF SIDE OF HEAD, NECK, TRUNK, AND UPPER EXTREMITIES.

Fig. 1.

Interior of Thorax, as when laid open, exposing Pericardium, Lungs, and large Blood-vessels at root of Neck.

a Larynx.
b Trachea.
c Lobus superior pulmonis.
d Lobus inferior pulmonis.
e Lobus medius pulmonis.
f Diaphragm.
g Processus xiphoideus sterni.
h Mediastinum anticum pleuræ.
i Pars diaphragmatica pleuræ.
k Pericardium.
  1. Vena cava superior.
  2. Ven. innominata sinistra.
  3. Ven. innominata dextra.
  4. Ven. subclavia.
  5. Ven. jugularis interna.
  6. Arcus aortæ.
  7. Arteria innominata.
  8. Art. carotis communis.
  9. Art. subclavia.
  10. Art. et ven. pericardiaco-phrenica.

Fig. 2.

Posterior surface of Lungs and Trachea, with their principal Arteries, Veins, and Nerves.

a Larynx.
b Trachea.
c Bronchus dexter.
d Bronchus sinister.
e Lobus superior pulmonis.
f Lobus inferior pulmonis.
g Lobus medius pulmonis.
h Right auricle with orifice of inferior vena cava.
i Left orifice.
k Right ventricle.
l Left ventricle.
  1. Pulmonary veins of right and left lungs.
  2. Vena magna cordis.
  3. Art. coronaria cordis dextra.
  4. Artt. pulmonales.
  5. Arcus aortæ.
  6. Art. innominata.
  7. Art. subclavia.
  8. Art. carotis communis.
  9. Vena jugularis interna.
  10. Vena cava superior.
  11. Nervus vagus v. pneumo-gastricus.
  12. Ramus recurrens nervi vagi v. nerv. laryngeus inferior.
  13. Rami trachealos nerv. recurrentis.
  14. Rami cardiaci nerv. recurrentis.
  15. Nerv. laryngeus superior.
  16. Ramus cardiacus nervi sympathici.
  17. Plexus cardiacus.

Fig. 3.

Distribution of Internal Maxillary and Labial or Facial Arteries and Veins on left side of Head.

a Os frontis.
b Ala magna ossis sphenoidei.
c Os maxillare superius.
d Inner wall of orbit.
e Malar bone.
f Inferior maxilla.
g Body of maxilla.
h M. pterygoideus externus.
i M. pterygoideus internus.
k Masseter.
l Orbicularis oris.
m Buccinator.
  1. Carotis communis sinistra.
  2. Vena jugularis interna.
  3. Ven. jugularis externa.

  4. Ven. labialis.
  5. Ven. facialis anterior.
  6. Ven. facialis posterior.
  7. Art. et ven. occipitalis.
  8. Art. et ven. auricularis posterior.
  9. Art. temporalis (superficialis).
  10. Art. maxillaris interna.
  11. Artt. and vv. temporales profundæ.
  12. Art. and ven. alveolaris inferior.
  13. Art. and ven. alveolaris posterior.
  14. Art. maxillaris externa v. facialis v. labialis.
  15. Art. coronaria labii inferioris.
  16. Art. coronaria labii superioris.
  17. Art. dorsalis nasi.
  18. Art. angularis.
  19. Art. ophthalmica and ven. ophthalmica cerebralis.
  20. Art. and ven. frontalis.

Fig. 4.

Principal Arteries and Veins of Neck, Thorax, and Arms, with deep Blood-vessels of Abdominal Cavity.

a Maxilla inferior.
b Os hyoides.
c Larynx.
d Glandula thyroidea.
e Trachea.
f Œsophagus.
g Clavicle.
h First rib.
i Lung.
k Heart.
l Pericardium.
m Right auricle.
n Left auricle.
o Right ventricle.
p Left ventricle.
q Diaphragm.
r Œsophagus.
s Kidney.
t Glandula suprarenalis.
u Ureter.
v Vesica urinaria.
w Intestinum rectum.
x Peritoneum.
y Muscl. quadratus lumborum.
z M. psoas.
α M. transversalis abdominis.
β M. iliacus internus.
γ Spermatic cord.
δ Muscl. sartorius.
ε Poupart's ligament.
ζ Muscl. pectoralis major.
η M. trapezius.
θ M. scalenus anticus.
ι M. deltoideus.
κ M. biceps flexor cubiti.
λ M. brachialis anticus.
μ M. triceps extensor.
ν M. supinator longus.
ξ M. flexor carpi ulnaris.
ο M. flexor pollicis longus.
π M. flexor digitorum communis profundus.
ρ M. pronator quadratus.
  1. Vena cava superior.
  2. Arteria aorta ascendens.
  3. Arteria pulmonalis.
  4. Arcus aortæ.
  5. Art. innominata.
  6. Art. carotis communis dextra.
  7. Art. subclavia dextra.
  8. Art. carotis communis sinistra.
  9. Art. subclavia sinistra.
  10. Vena innominata sinistra.
  11. Vena innominata dextra.
  12. Vena jugularis interna.
  13. Ven. jugularis externa.
  14. Ven. subclavia.
  15. Ven. thyroidea superior.
  16. Venæ subcutaneæ colli.
  17. Ven. thyroidea superior.
  18. Ven. labialis.

  19. Ven. cephalica posterior v. interna.
  20. Arteria facialis v. labialis.
  21. Vena facialis anterior.
  22. Arteriæ pulmonales.
  23. Venæ pulmonales.
  24. Ramus anterior v. descendens art. et ven. coronariæ cordis sinistræ.
  25. Art. et ven. coronaria cordis dextra.
  26. Aorta descendens abdominalis.
  27. Artt. phrenicæ inferiores.
  28. Art. axis cœliaca.
  29. Art. mesenterica superior.
  30. Artt. spermaticæ internæ.
  31. Art. mesenterica inferior.
  32. Art. hæmorrhoidalis interna.
  33. Art. and vena renalis.
  34. Art. iliaca communis.
  35. Art. iliaca interna.
  36. Art. iliaca externa.
  37. Art. and ven. circumflexa ilium.
  38. Art. and ven. ilio-lumbalis.
  39. Vena cava inferior.
  40. Venæ hepaticæ.
  41. Vena renalis.
  42. Ven. spermatica interna.
  43. Ven. iliaca communis.
  44. Ven. iliaca interna.
  45. Ven. iliaca externa.
  46. Art. and ven. sacra media.
  47. Nerv. inguino cutaneus.
  48. Nerv. ilio-lumbalis.
  49. Art. axillaris.
  50. Ven. axillaris.
  51. Ven. cephalica brachii.
  52. Ven. basilica.
  53. Ven. mediana.
  54. Art. brachialis.
  55. Bifurcation of brachial artery.
  56. Art. radialis.
  57. Art. ulnaris.
  58. Art. interossea communis.
  59. Art. interossea interna.
  60. Art. recurrens radialis.
  61. Art. recurrens ulnaris.
  62. Arcus palmaris sublimis.
  63. Ramus superficialis volæ arteriæ radialis.

Fig. 5.

Section of Spinal Column, with Vena Azygos, right Intercostal Vessels, Aorta, and Venæ Cavæ.

a Costa (1), first rib.
b Costa (12), last rib.
c Vertebra lumbalis (1), first lumbar vertebra.
d Vena cava superior.
e Vena cava inferior.
f Aorta descendens thoracica.
g Arcus aortæ.
  1. Vena azygos.
  2. Termination of vena azygos in vena cava superior.
  3. Artt. and vv. lumbales.
  4. Artt. and vv. intercostales.

Fig. 6.

Posterior inverted view of deep Vessels of Thorax, Vena Azygos, Aorta, and Venæ Cavæ.

a First rib.
b Sternum.
c Costal cartilages.
d Vena cava superior.
e Vena cava inferior.
f Aorta ascendens.
g Arcus aortæ.
h Aorta descendens thoracica.
i Aorta descendens abdominalis.
  1. Vena azygos.
  2. Vena demiazygos.
  3. Artt. and vv. intercostales.
  4. Artt. and vv. lumbales.

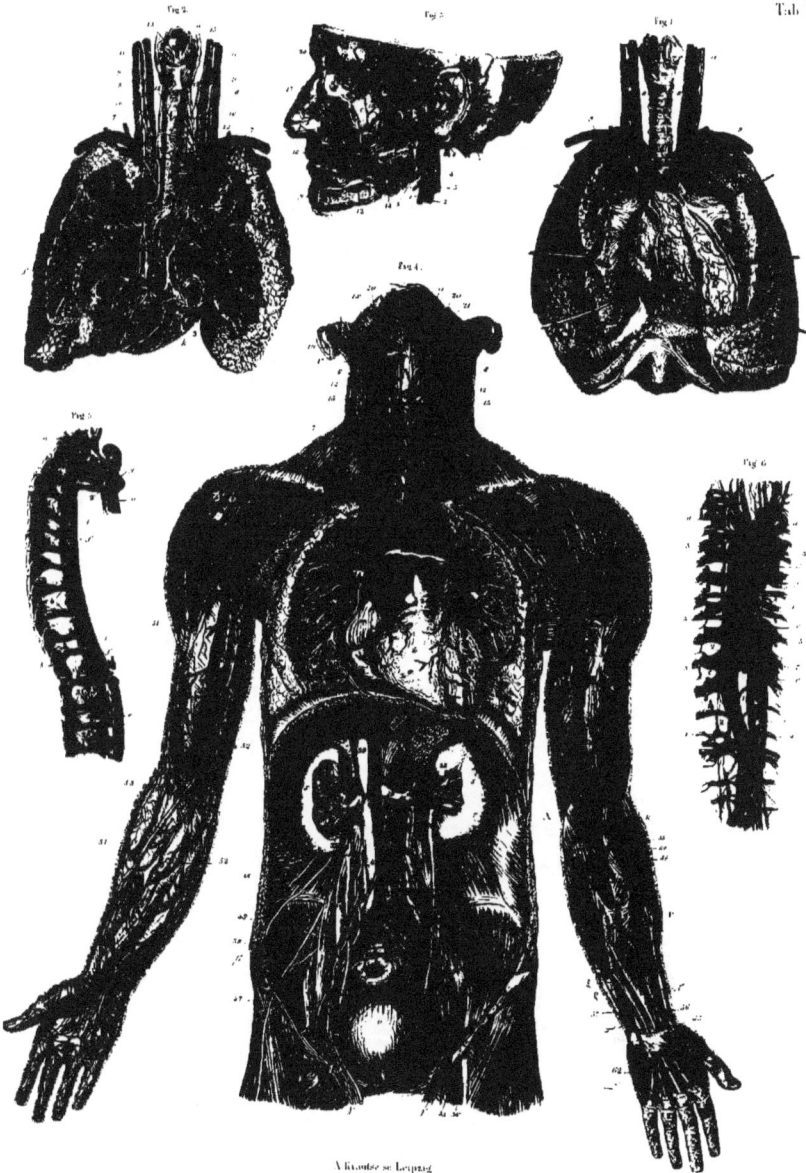

Tab. XIX.

Fig. 2. Fig. 3. Fig. 1.

Fig. 4.

Fig. 5. Fig. 6.

A lithuanie se Leipzig

PLATE XX.

BLOOD-VESSELS OF PERINEAL REGIONS (MALE AND FEMALE), FEMALE INTERNAL ORGANS OF GENERATION, INGUINAL REGION, BACK OF SHOULDER AND FOREARM, AND PELVIS.

Fig. 1.

Arteries on posterior surface of Shoulder and Scapula.

a   Clavicula.
b   Acromion.
c   Spina scapulæ.
d   Fossa supraspinata.
e   Fossa infraspinata.
f   Angulus inferior scapulæ.
g   Os humeri.
h   Tuberculum majus humeri.
i   Musculus levator anguli scapulæ.
k   Macl. rhomboideus minor.
l   M. rhomboideus major.
m   M. infraspinatus.
n   M. teres minor.
o   M. teres major.
p   M. deltoideus.
q   M. trapezius.
r   M. serratus anticus major.
s   M. latissimus dorsi.
t   Caput longum m. tricipitis extensoris.

   1. Arteria subclavia.
   2. Art. cervicalis superficialis.
   3. Art. transversalis colli.
   4. Art. dorsalis scapulæ.
   5. Art. transversalis scapulæ.
   6. Art. acromialis.
   7. Ramus infraspinatus art. transversalis scapulæ.
   8. Art. circumflexa scapulæ.
   9. Ramus descendens art. subscapularis.
   10. Art. circumflexa humeri posterior.

Fig. 2.

Deep Arteries on outer side of Arm, and posterior surface of Forearm, Wrist, and Hand.

a   Os humeri.
b   Condylus externus humeri.
c   Olecranon ulnæ.
d   Ulna.
e   Capitulum radii.
f   Radius.
g   Processus styloideus radii.
h   Carpus.
i   Metacarpus.
k   Ligamentum interosseum.
l   Macl. brachialis anticus.
m   Mm. interossei dorsales v. externi.

   1. Arteria superior profunda brachii.
   2. Art. recurrens interossea.
   3. Art. interossea perforans.
   4. Rami perforantes art. interosseæ internæ.
   5. Art interossea perforans inferior.
   6. Ramus dorsalis arteriæ ulnaris.
   7. Art. radialis.
   8. Arcus dorsalis carpi.
   9. Art. dorsalis radialis.
   10. Art. dorsalis ulnaris (5).
   11. Artt. interosseæ metacarpi dorsales.
   12. Artt. digitales dorsales.
   13. Art. dorsalis ulnaris pollicis.
   14. Art. dorsalis radialis indicis.

Fig. 3.

Arteries and Veins on anterior surface of Inguinal Region—left side.

a   Spina ilii anterior superior.
b   Musculus obliquus abdominis externus.
c   Macl. obliquus abdominis internus.
d   M. transversalis abdominis.
e   M. rectus abdominis.
f   External abdominal ring.
g   Spermatic cord.
h   M. sartorius.
i   M. iliacus internus.
k   M. psoas major.
l   M. pectineus.
m   M. adductor longus.
n   M. rectus femoris.
o   M. tensor fasciæ latæ.
p   Penis with art. and ven. dorsalis penis
q   Inguinal canal.

   1. Arteria femoralis.
   2. Art. profunda femoris.
   3. Art. circumflexa femoris externa.
   4. Art. and ven. circumflexa ilii.
   5. Art. and ven. epigastrica (inferior).
   6. Vena femoralis.
   7. Ven. profunda femoris.
   8. Ven. saphena.
   9. Ven. circumflexa femoris externa

Fig. 4.

Arteries of Ano-Perineal Region in the Female subject.

a   Tuber ossis ischii.
b   Ramus ascendens ossis ischii.
c   Great sacro-sciatic ligament.
d   Clitoris.
e   Orificium urethræ.
f   Ostium.
g   Anus.
h   Lesser sacro-sciatic ligament
i   M. constrictor vaginæ.
k   Muscular fibres of constrictor vaginæ lining upper surface of urethra.
l   Decussation of fibres of mm. constrictor vaginæ and sphincter ani.
m   M. sphincter ani externus.
n   M. ischio-cavernosus.
o   Mm. transversal perinæi.
p   M. levator ani.
q   M. gracilis.
r   M. adductor magnus
s   M. glutæus maximus

   1. Art. pudenda interna.
   2. Anastomosis between internal pubic, vaginal, and clitorideal branches.
   3. Artt. hæmorrhoidales externæ.
   4. Art. transversa perinæi.
   5. Art. vaginalis.
   6. Art. clitoridea.
   7. Art. dorsalis clitoridis.
   8. Art. profunda clitoridea.

Fig. 5.

Arteries and Veins of Ano-Perineal Region in the Male subject.

a   Tuber ossis ischii.
b   Ramus ascendens ossis ischii.
c   Great sacro-sciatic ligament.
d   Lesser sacro-sciatic ligament
e   Bulb of urethra.
f   Corpus cavernosum penis

g   Anus.
h   Os coccygis.
i   M. acceleratores urinæ.
k   M. erector penis.
l   M. transversus perinæi
m   M. sphincter ani externus.
n   M. levator ani.
o   M. adductor longus.
p   M. gracilis.
q   M. adductor magnus.
r   M. glutæus maximus.

   1. Art. pudenda interna.
   2. Artt. hæmorrhoidales externæ.
   3. Art. transversa perinæi.
   4. Art. bulbo-urethralis.
   5. Art. perinæi superficialis.

Fig. 6.

Arteries of Pelvis in the Male subject.

a   Last lumbar vertebra.
b   Os sacrum.
c   Crest of ilium.
d   Musculus psoas (magnus)
e   Macl. iliacus internus.
f   M. transversalis abdominis.
g   M. rectus abdominis.
h   Ureter.
i   Vesica urinaria.
k   Rectum.
l   Vas deferens.

   1. Aorta descendens abdominalis.
   2. Arteria mesenterica inferior.
   3. Art. hæmorrhoidalis interna.
   4. Art. colica sinistra.
   5. Art. sacra media.
   6. Art. spermatica interna.
   7. Art. iliaca communis.
   8. Art. iliaca externa.
   9. Art. iliaca interna.
   10. Art. circumflexa ilii.
   11. Art. epigastrica (inferior).
   12. Art. ilio-lumbalis.

Fig. 7.

Arteries of Pelvis and internal Genital Organs in the Female subject.

a   Os sacrum.
b   Crest of ilium.
c   Spina ilii anterior superior.
d   Macl. psoas magnus.
e   M. iliacus internus
f   Intestinum rectum.
g   Uterus.
h   Ligamentum uteri latum.
i   Ovarium with ligamentum ovarii.
k   Tuba Fallopii.
l   Ligamentum uteri latum.

   1. Aorta descendens abdominalis.
   2. Art. sacra media.
   3. Art. spermatica interna.
   4. Art. iliaca interna.
   5. Art. iliaca externa.
   6. Art. iliaca interna.
   7. Art. uterina.
   8. Art. hæmorrhoidalis media.
   9. Art. circumflexa ilii.

Fig. 3

Fig. 2

Fig. 1

Fig. 4.

Fig. 5.

Fig. 7.

Fig. 6.

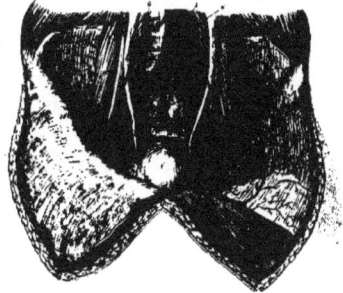

A. Brandus se Leipzig

PLATE XXI.

ARTERIES OF PELVIS AND LOWER EXTREMITIES.

Fig. 1.

Arteries on anterior surface of Pelvis, Thigh, Leg, and Foot of right extremity.

a Crista ilii.
b Spina ilii anterior.
c Trochanter major.
d Symphysis cedum pubis.
e Ramus horizontalis ossis pubis.
f Patella.
g Tuberositas tibiæ.
h Tibia.
i Malleolus internus.
k Malleolus externus.
l Musculus psoas.
m Muscl. iliacus internus.
n M. pectineus.
o M. glutæus medius.
p M. rectus femoris.
q M. tendo communis extensorius.
r M. vastus externus.
s M. vastus internus.
t M. cruralis.
u M. adductor longus.
v M. gracilis.
w M. adductor magnus.
x Ligamentum patellæ.
y M. tibialis anticus.
z M. extensor pollicis pedis longus.
α M. extensor digitorum communis longus.
β M. peronæus tertius.
γ M. soleus.
δ M. gastrocnemius.
ε M. extensor pollicis pedis brevis.
ζ M. extensor digitorum communis brevis.
 1. Arteria iliaca communis dextra.
 2. Art. iliaca interna.
 3. Art. sacra lateralis.
 4. Art. glutæa (superior).
 5. Art. ischiadica v. glutæa (inferior).
 6. Art. iliaca externa.
 7. Art. epigastrica (inferior).
 8. Art. circumflexa ilii.
 9. Art. femoralis.
 10. Art. profunda femoris.
 11. Art. circumflexa femoris externa.
 12. Entrance of the femoral artery into Hunter's canal.
 13. Art. tibialis antica.
 14. Art. recurrens tibialis.
 15. Art. dorsalis pedis.
 16. Art. malleolaris externa.
 17. Art. malleolaris interna.
 18. Art. tarsea externa.
 19. Art. tarsea interna.
 20. Artt. interosseæ metatarsi dorsales.

Fig. 2.

Arteries on internal surface of Pelvis, Thigh, and Knee of the right extremity.

a Fourth lumbar vertebra.
b Fifth lumbar vertebra.
c Spinal canal.
d Os sacrum.
e Os coccygis.
f Linea arcuata interna.
g Symphysis pubis.
h Crista ilii.
i Spina ilii anterior superior.
k Lesser sacro-sciatic ligament.
l Rectum.
m Musculus iliacus internus.
n Muscl. psoas major.
o M. pyriformis.
p M. obturator internus.
q M. levator ani.
r M. sartorius.
s M. vastus internus.
t M. rectus femoris.
u M. adductor magnus.
v M. semimembranosus.
w M. semitendinosus.
x Tendo m. gracilis.
y M. gastrocnemius (internus).
z M. soleus.
 1. Arteria iliaca communis dextra.
 2. Art. iliaca interna.
 3. Art. iliaca externa.
 4. Art. ilio-lumbalis.
 5. Art. obturatoria.
 6. Art. sacra lateralis.
 7. Art. glutæa (superior).
 8. Art ischiadica v. glutæa inferior.
 9. Art. pudenda interna.
 10. Artt. hæmorrhoidales mediæ.
 11. Art. vesicalis.
 12. Art. circumflexa ilii.
 13. Art. femoralis.
 14. Art. profunda femoris.
 15. Art. circumflexa femoris interna.
 16. Art. perforans (1) arteriæ profundæ femoris.
 17. Art. perforans (2) arteriæ profundæ femoris.
 18. Art. perforans (3) arteriæ profundæ femoris.
 19. Art. femoralis, in Hunter's canal.
 20. Art. anastomotica magna.
 21. Art. poplitea.
 22. Art. articularis genu inferior interna.

Fig. 3.

Arteries on posterior surface of Pelvis, Thigh, and Leg of the right extremity.

a Os sacrum.
b Os coccygis.
c Crista ossis ilii.
d Os ilium.
e Trochanter major.
f Tuber ischii.
g Ramus ascendens ossis ischii.
h Ramus descendens ossis pubis.
i Great sacro-sciatic ligament.
k Fossa poplitea.
l Capitulum fibulæ.
m Fibula.
n Malleolus externus.
o Malleolus internus.
p Musculus glutæus medius.
q Musc. pyriformis.
r M. gemellus superior.
s M. obturator internus.
t M. gemellus inferior.
u M. quadratus femoris.
v M. adductor femoris magnus.
w Origin of biceps and semitendinosus.
x M. semimembranosus.
y Caput breve m. bicipitis.
z M. vastus externus.
α M. gracilis.
β M. popliteus.
γ Heads of m. gastrocnemius.
δ M. peronæus longus.
ε M. peronæus brevis.
ζ M. flexor longus pollicis pedis.
η M. tibialis posticus.
ϑ M. flexor digitorum longus.
ι Tendo Achillis.
κ M. soleus.
 1. Arteria glutæa superior.
 2. Art. glutæa inferior.
 3. Art. pudenda interna.
 4. Ramus inferior art. circumflexæ femoris internæ.
 5. Art. profunda femoris.
 6. Art. perforans (1), arteriæ profundæ femoris.
 7. Art. perforans (2), arteriæ profundæ femoris.
 8. Art. perforans (3), arteriæ profundæ femoris.
 9. Art. poplitea.
 10. Art. articularis genu superior interna.
 11. Art. articularis genu superior externa.
 12. Art. articularis genu inferior interna.
 13. Art. articularis genu inferior externa.
 14. Art. tibialis antica.
 15. Art. fibularis v. peronæa.
 16. Art. tibialis postica.
 17. Art. malleolaris posterior externa.

Fig. 4.

Arteries on dorsal surface of Right Foot.

a Astragalus.
b Os calcis.
c Os naviculare.
d Tuber ossis metatarsi (5).
 1. Arteria dorsalis pedis.
 2. Art. tarsea externa.
 3. Art. tarsea interna.
 4. Art. metatarsea.
 5. Artt. interosseæ metatarsi dorsales.
 6. Artt. digitales pedis dorsales.
 7. Art. interossea dorsalis with the external and internal dorsal branches to the great toe.
 8. Communicating branch to deep plantar arch.

Fig. 5.

Plantar arch of Arteries in sole of Right Foot.

a Tuber os calcis.
b Tuberositas ossis metatarsi (5).
c Capitulum ossis metatarsi (1).
d M. flexor digitorum pedis communis brevis v. perforatus.
e M. abductor pollicis pedis.
f M. flexor brevis pollicis pedis.
g M. flexor longus pollicis pedis.
h M. flexor digitorum pedis communis longus v. perforans.
i Massa carnea vel musc. accessorius.
k M. abductor digiti pedis (5).
l M. flexor brevis digiti pedis (5).
m M. transversalis pedis.
 1. Art. tibialis postica.
 2. Art. plantaris externa.
 3. Branches of art. plantaris interna.
 4. Art. plantaris digiti (5) externa.
 5. Communicating branch of deep plantar arch.
 6. Art. plantaris pollicis pedis.
 7. Artt. interosseæ plantares.

Fig. 6.

Deep arteries in sole of Right Foot.

a Tuber os calcis.
b M. adductor pollicis pedis.
c Mm. interossei pedis plantares.
d M. flexor brevis digiti (5).
 1. Art. tibialis postica.
 2. Art. plantaris externa.
 3. Art. plantaris interna.
 4. Art. tibialis plantaris pollicis pedis.
 5. Perforating branches.
 6. Arcus plantaris profundus.
 7. Artt. interosseæ plantares.
 8. Artt. digitales pedis plantares.
 9. Art. plantaris digiti (5) externa.

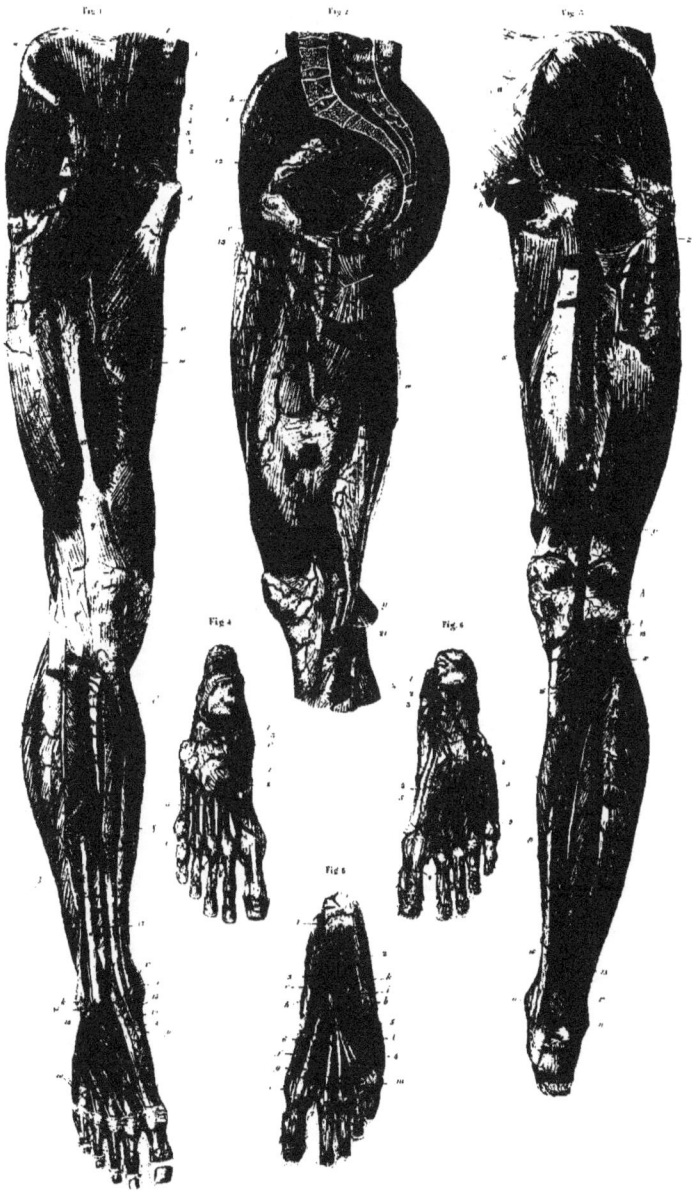

Tab. XXI.

PLATE XXII.

THE THORACIC DUCT, LYMPHATIC VESSELS, AND GLANDS.

Fig. 1.

Thoracic Duct, Lymphatics of Trunk, Pelvis, upper part of Thighs, and superior extremities.

a Larynx.
b Trachea.
c Crista ilii.
d Symphysis pubis.
e Spina ilii anterior superior.
f Clavicle.
g First rib.
h Macl. frontalis.
i M. orbicularis palpebrarum.
k M. masseter.
l Mm. zygomatici.
m M. orbicularis oris.
n M. quadratus menti.
o M. triangularis menti.
p Diaphragm.
q M. quadratus lumborum.
r M. psoas major.
s M. iliacus internus.
t M. deltoideus.
u M. biceps humeri.
v M. pectineus.
w M. adductor longus.
x M. sartorius.
y M. rectus femoris.
z M. vastus internus.
  1. Vena cava superior.
  2. Ven. innominata dextra.
  3. Ven. innominata sinistra.
  4. Ven. jugularis interna.
  5. Ven. jugularis externa.
  6. Ven. subclavia.
  7. Art. subclavia.
  8. Art. carotis communis.
  9. Art. facialis v. maxillaris externa.
  10. Ven. facialis anterior.

11. Art. and ven. temporalis.
12. Art. axillaris.
13. Art. brachialis.
14. Vena cephalica brachii.
15. Ven. basilica.
16. Ven. mediana.
17. Art. and ven. ulnaris.
18. Art. and ven. radialis.
19. Art. and ven. interossea interna.
20. Aorta descendens abdominalis.
21. Art. iliaca communis.
22. Art. femoralis.
23. Vena cava inferior.
24. Vena azygos.
25. Vena saphena magna.
26. Receptaculum chyli, the commencement of thoracic duct.
27. Termination of thoracic duct in the junction of the left internal jugular and subclavian veins.
28. Right thoracic duct.
29. Plexus lymphaticus cervicalis v. jugularis.
30. Plexus lymphaticus axillaris.
31. Glandulæ brachiales.
32. Superficial lymphatics of upper extremity.
33. Deep lymphatics of upper extremity.
34. Plexus lymphatici intercostales.
35. Plexus lymphaticus lumbalis.
36. Plexus lymphatici iliaci.
37. Plexus inguinalis superficialis.
38. Plexus inguinalis profundus.
39. Plexus saphenus.

Fig. 2.

Lymphatic Vessels and Glands of Axilla (left side).

a Arteria axillaris.
b Vena axillaris.

c Art. and ven. brachialis.
d Vena jugularis interna.
e Art. carotis communis.
f Ven. jugularis externa.
g Trachea.
h Macl. pectoralis major.
i M. latissimus dorsi.
  1. Ductus thoracicus.
  2. Superficial lymphatics.
  3. Deep lymphatics.
  4. Glandula brachialis.
  5. Glandulæ axillares in plexus lymphaticus axillaris.
  6. Vasa lymphatica thoracica.
  7. Vasa lymphatica jugularia.

Fig. 3.

Lymphatic Vessels and Glands in interior of Pelvis (right side).

a Aorta abdominalis.
b Vena cava inferior.
c Arteria iliaca communis.
d Vena iliaca communis.
e Art. iliaca externa.
f Ven. iliaca externa.
g Ven. iliaca interna.
h Art. and ven. sacra media.
i Art. and ven. obturatoria.
k Art. and ven. epigastrica.
l Art. and ven. pudenda interna.
m Art. and ven. circumflexa ilii.
n Macl. pyriformis.
o M. obturator internus.
  1. Glandulæ iliacæ internæ.
  2. Vasa lymphatica pelvina profunda.

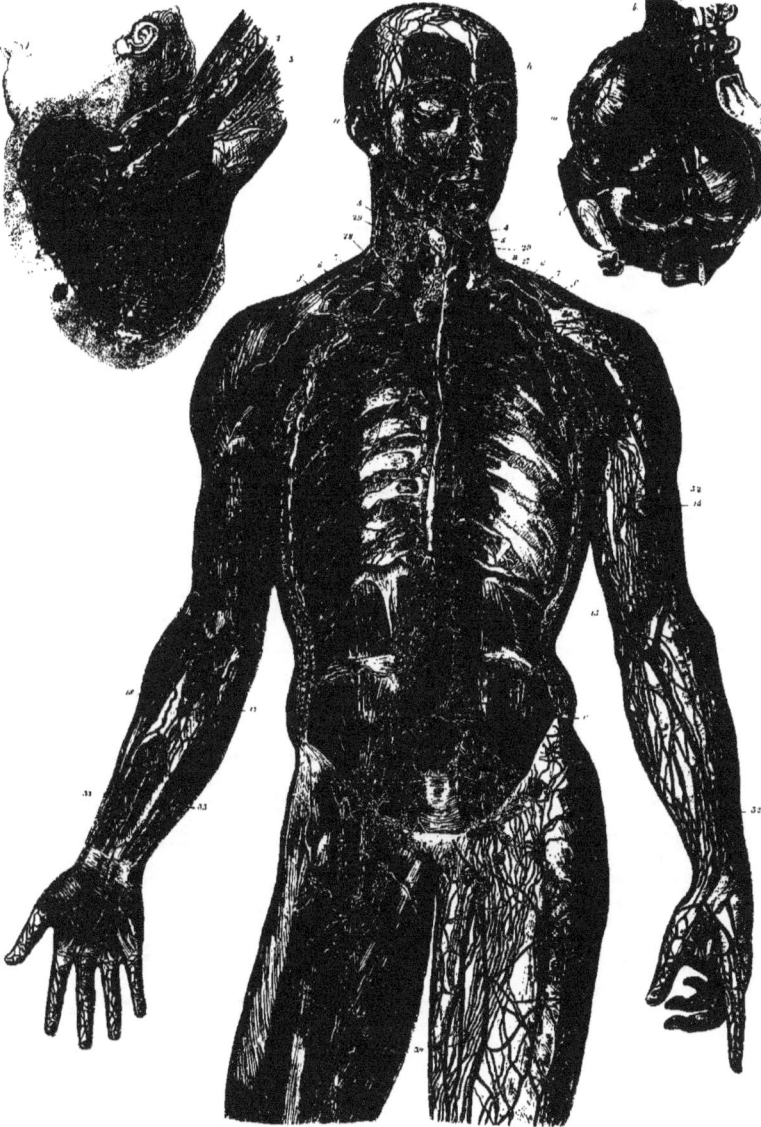

PLATE XXIII.

THE BRAIN: CEREBRUM, CEREBELLUM, AND MEDULLA OBLONGATA.

Fig. 1.

The under surface or Base of the Brain, with origin of Cerebral Nerves.

I. Os frontis.
II. Os occipitis.
III. Os temporum.
IV. Falx cerebri.
V. Falx cerebelli.
VI. Sinus longitudinalis superior.
VII. Sinus occipitalis.
  A. Lobus anterior cerebri.
  B. Lobus medius v. inferior cerebri.
  C. Lobus posterior cerebri.
  D. Cerebellum.
  E. Medulla oblongata.
  a. Convolutiones.
  b. Sulci.
  c. Fissure of Sylvius.
  d. Corpus pyramidale medullæ oblongatæ.
  e. Corpus olivare medullæ oblongatæ.
  f. Pons Varolii v. commissura cerebelli.
  g. Crus cerebelli.
  h. Crus cerebri.
  i. Tractus opticus.
  k. Commissura nervorum opticorum.
  l. Fundus ventriculi tertii v. substantia perforata media.
  m. Corpora mammillaria v. candicantia.
  n. Tuber cinereum.
  o. Infundibulum.
  p. Glandula pituitaria.
  q. Lamina cribrosa v. substantia perforata anterior.
  r. Root of olfactory nerve.
  s. Olfactory tract.
  t. Convolution of hippocampus.
  u. Uncus.
  v. Horizontal sulcus (Reilii) between cerebrum and cerebellum.
  w. Pyramis vermis.
  x. Lobulus quadrangularis.
  y. Lobulus semilunaris.
  z. Lobulus inferior anterior.
  α. Lobulus tener.
  β. Lobulus biventer.
  γ. Tonsilla v. lobulus spiralis.
  δ. Flocculus.
    1. Nerv. olfactorius (bulbus cinereus).
    2. Nerv. opticus.
    3. Nerv. oculo-motorius (third nerve).
    4. Nerv. trochlearis v. patheticus (fourth nerve).
    5. Nerv. trigeminus (fifth nerve).
    6. Nerv. abducens (sixth nerve).
    7. Nerv. facialis v. communicans faciei.
    8. Nerv. acousticus v. auditorius, portio mollis of ninth nerve.
    9. Nerv. glosso-pharyngeus (eighth nerve)
    10. Nerv. vagus v. pneumo-gastricus (eighth nerve).
    11. Nerv. accessorius Willisii, or spinal accessory nerve (eighth nerve).
    12. Nerv. hypoglossus (ninth nerve).

Fig. 2.

Central surface of Cerebrum, shown by making a horizontal section through its centre, exposing the interior of the two lateral, third, and fifth Ventricles.

  A. Lobus anterior cerebri.
  B. Lobus medius cerebri.
  C. Lobus posterior cerebri.
  D. Cerebellum (upper surface).
  E. Substantia corticalis v. cinerea.
  F. Substantia alba v. medullaria.
  I. Ventriculus lateralis dexter, cella lateralis.

II. Ventriculus lateralis, sinister cella lateralis.
III. Ventriculus tertius.
IV. Cornu anterius ventriculi lateralis.
V. Cornu posterius ventriculi lateralis.
VI. Cornu inferius ventriculi lateralis.
  a. Corpus callosum and raphe.
  b. Genu corporis callosi.
  c. Lyra.
  d. Septum lucidum and fifth ventricle.
  e. Fornix.
  f. Crura fornicis anteriora.
  g. Crura fornicis posteriora.
  h. Pes hippocampi major and corpus fimbriatum.
  i. Pes hippocampi minor.
  k. Corpus striatum.
  l. Thalamus nervi optici.
  m. Stria cornea.
  n. Commissura mollis ventriculi tertii.
  o. Commissura anterior ventriculi tertii.
  p. Commissura posterior ventriculi tertii.
  q. Aditus ad infundibulum.
  r. Aditus ad aquæductum Sylvii.
  s. Glandula pinealis.
  t. Crura glandulæ pinealis.
  u. Corpora quadrigemina.
  v. Plexus choroideus.

Fig. 3.

The inferior surface of Cerebellum, the Medulla Oblongata being removed.

  A. Pons Varolii.
  B. Crura cerebelli.
  C. Lateral lobes of cerebellum.
  D. Inferior vermiform appendix of cerebellum.
  a. Upper border of cerebellum.
  b. Lower border of cerebellum.
  c. Horizontal cleft.
  d. Superior anterior lobule.
  e. Inferior posterior lobule.
  f. Lobulus gracilis.
  g. Lobulus biventer.
  h. Tonsilla.
  i. Flocculus.
  k. Pyramis vermis.
  l. Nodulus Malacarne.

Fig. 4.

Posterior border and under surface of Cerebellum.

  A. Lateral lobe.
  B. Superior vermiform appendix.
  C. Inferior vermiform appendix.
  D. Spinal marrow.
  E. Horizontal cleft.
  a. Monticulus cerebelli.
  b. Folium cacuminis.
  c. Lobulus quadrangularis v. superior anterior.
  d. Tonsilla.
  e. Lobulus biventer.
  f. Lobulus gracilis.
  g. Lobulus semilunaris (inferior posterior).
  h. Pyramis vermis.
  i. Fasciculus lateralis medullæ spinalis.
  k. Fasciculus cuneatus medullæ spinalis.
  l. Fasciculus gracilis medullæ spinalis.
  m. Posterior roots of spinal nerves.

Fig. 5.

Pons Varolii and Medulla Oblongata—anterior surface.

  A. Pons Varolii.
  B. Medulla oblongata.

  a. Crus cerebelli.
  b. Corpus restiforme.
  c. Corpus olivare.
  d. Corpus pyramidale.
  e. Decussatio pyramidum.
  f. Fibræ arciformes.
  g. Fissura longitudinalis anterior.
    1. Nervus accessorius.
    2. Nerv. vagus v. pneumogastricus.
    3. Nerv. glossopharyngeus.
    4. Nerv. acousticus v. auditorius, portio mollis of seventh pair.
    5. Nerv. facialis, portio dura of seventh pair.
    6. Portio minor of portio dura.

Fig. 6.

Pons Varolii and Medulla Oblongata—anterior surface, showing internal fibres.

  A. Pons Varolii.
  B. Medulla oblongata.
  a. Crus cerebelli.
  b. Corpus restiforme.
  c. Corpus olivare.
  d. Corpus pyramidale.
  e. Decussatio pyramidum.
  f. Fibræ transversæ.
  g. Fibræ arciformes.
  h. Fissura longitudinalis anterior.

Fig. 7.

Pons Varolii and right lateral Lobe of Cerebellum, showing transverse fibres of Pons and Crus Cerebelli continuous with Arbor Vitæ.

  A. Pons Varolii.
  B. Medulla oblongata.
  C. Cerebellum.
  a. Crus cerebelli ad pontem.
  b. Crus cerebelli ad medullam oblongatam v. corpus restiforme.
  c. Corpus olivare.
  d. Corpus pyramidale.
  e. Decussatio pyramidum.
  f. Fibræ transversæ pontis.
  g. Fibræ longitudinales pyramidum.
  h. Arbor vitæ cerebelli.

Fig. 8.

Vertical Section of Pons Varolii and lateral Lobe of Cerebellum, showing internal arrangement of fibres.

  A. Pons Varolii.
  B. Medulla oblongata.
  C. Cerebellum, left lobe.
  a. Left pyramidal body.
  b. Fasciculus olivaris sinister.
  c. Nucleus olivæ.
  d. Corpus restiforme sinistrum.
  e. Pedunculus cerebelli (crus cerebelli ad medullam oblongatam.)
  f. Crus cerebelli ad corpora quadrigemina.
  g. Corpora quadrigemina.
  h. Corpus mammillare.
  i. Fibræ horizontales pyramidis dextræ.
  k. Fibræ longitudinales superficiales et Profundæ pontis (pyramidis).
  m. Fibræ longitudinales pontis profundæ et Laqueus fasciculi olivaris.
  o. Pedunculus cerebri.
  p. Fibræ ad pedunculum cerebri.

Tab. XXIII.

Fig. 1.

Fig. 5.

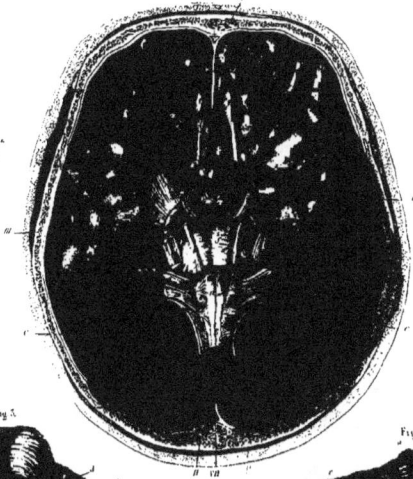

Fig. 6.

Fig. 3.

Fig. 4.

Fig. 2.

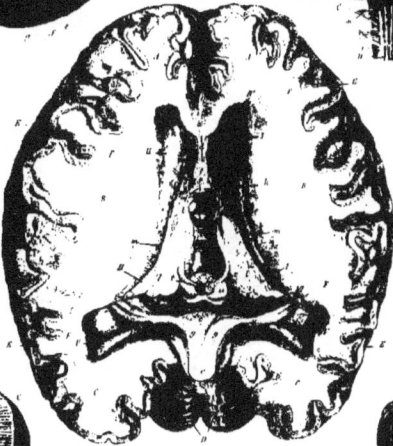

Fig. 8.

Fig. 7.

PLATE XXIV.

BASE AND INTERIOR OF BRAIN, WITH ORIGINS OF NERVES AND BLOOD-VESSELS.

Fig. 1.

Base of Brain, showing origins of Nerves and the Arteries.

A. Lobus anterior cerebri.
B. Lobus media cerebri.
C. Lobus posterior.
D. Cerebellum (lateral lobes of).
E. Medulla oblongata.
a. Fissura Sylvii.
b. Fissura longitudinalis cerebri.
c. Commissura nervorum opticorum.
d. Tuber cinereum.
e. Corpora mammillaria v. candicantia.
f. Tractus opticus.
g. Pons Varolii.
h. Crus cerebelli ad pontem.
i. Corpus pyramidale.
k. Corpus olivare.
  1. Tractus olfactorius (first pair of nerves).
  2. Nervus opticus (second pair of nerves).
  3. Nerv. oculomotorius (third pair of nerves).
  4. Nerv. trochlearis (fourth pair of nerves).
  5. Nerv. trigeminus (fifth pair of nerves).
  6. Nerv. abducens (sixth pair of nerves).
  7. Nerv. facialis, portio mollis of seventh nerve.
  8. Nerv. acousticus, portio mollis of seventh nerve.
  9. Nerv. glossopharyngeus of eighth pair of nerves.
 10. Nerv. vagus of eighth pair of nerves.
 11. Nerv. lingualis or ninth pair of nerves.
 12. Arteria vertebralis.
 13. Art. basilaris.
 14. Artt. spinales anteriores.
 15. Artt. cerebelli inferiores posteriores.
 16. Artt. cerebelli inferiores anteriores.
 17. Artt. cerebelli superioris.
 18. Art. cerebri profunda.
 19. Rami communicantes (forming with anterior cerebral, internal carotid, and posterior or deep cerebral arteries, the circle of Willis).
 20. Art. carotis interna.
 21. Art. fossae Sylvii.
 22. Art. choroidea.
 23. Art. corporis callosi.

Fig. 2.

Section of Brain, exposing interior of Ventricles.

A. Portion of middle lobe.
B. Cerebellum.
C. Medulla oblongata.
D. Ventriculus lateralis.
E. Ventriculus tertius.
F. Ventriculus quartus.
a. Corpus striatum.
b. Thalamus nervi optici.
c. Stria cornea.

d. Crura glandulae pinealis.
e. Glandula pinealis.
f. Corpora quadrigemina.
g. Aditus ad aquaeductum Sylvii.
h. Aditus ad infundibulum.
i. Commissura anterior.
k. Fornix.
l. Crus cerebelli ad corpora quadrigemina.
m. Floor of fourth ventricle.
n. Sinus rhomboideus.
o. Corpus restiforme v. crus cerebelli ad medullam oblongatam.
p. Arbor vitae cerebelli.

Fig. 3.

Vertical longitudinal section of Brain, Cerebrum, and Cerebellum, through the centre.

  I. Os frontis et sinus frontalis.
 II. Crista galli.
III. Lamina perpendicularis ossis ethmoidei.
IV. Corpus ossis sphenoidei.
 V. Processus clinoidei posteriores.
VI. Sella turcica.
VII. Sinus sphenoidalis.
VIII. Pars basilaris ossis occipitis et fossa pro medulla oblongata.
IX. Pars occipitalis ossis occipitis.
 X. Vomer.
XI. Roof of pharynx.
XII. Tentorium cerebelli enclosing straight sinus.

A. Lobus anterior cerebri.
B. Lobus medius cerebri.
C. Lobus posterior cerebri.
D. Cerebellum (arbor vitae).
E. Medulla oblongata.
a. Convolutions of cerebrum.
b. Sulci.
c. Corpus callosum.
d. Genu corporis callosi.
e. Splenium corporis callosi.
f. Septum lucidum.
g. Fornix.
h. Crus anterius v. columna fornicis.
i. Foramen Monroi.
k. Thalamus nervi optici.
l. Commissura anterior.
m. Commissura mollis.
n. Commissura posterior.
o. Glandula pinealis.
p. Pedunculus v. crus glandulae pinealis.
q. Corpora quadrigemina.
r. Pons Varolii.
s. Aquaeductus Sylvii.
t. Tuber cinereum.
u. Infundibulum.
v. Glandula pituitaria.
w. Commissura nervorum opticorum.
x. Nervus opticus.
y. Ventriculus quartus.
z. Corpus mammillare v. candicans.

e. Valvula cerebelli anterior.
β. Art. corporis callosi.

Fig. 4.

Posterior View of Spinal Marrow (Medulla Spinalis), with anterior Roots of Spinal Nerves, the Spinal Sheath being divided and everted.

A. Medulla oblongata.
B. Pars cervicalis medullae spinalis.
C. Pars thoracica medullae spinalis.
D. Pars lumbalis medullae spinalis.
E. Ventriculus quartus.
a. Cervical expansion of medulla spinalis.
b. Lumbar expansion of medulla spinalis.
c. Bulbous or ganglionic expansion of medulla spinalis.
d. Filum terminale.
e. Fissura mediana posterior.
f. Ligamentum denticulatum.
  2—8) Nervi cervicales (2—8).
  9—20) Nervi thoracici (12).
 21—25) Nervi lumbales (5).
 26—30) Nervi sacrales (5).
   31) Nervi coccygei.
   32) Nervus accessorius spinalis in the foramen jugulare internum.
   33) Nervus vagus in the foramen jugulare internum.
   34) Nervus glossopharyngeus in the foramen jugulare internum.

Fig. 5.

Anterior view of Spinal Marrow (Medulla Spinalis), with posterior Roots of Spinal Nerves, the Spinal Sheath being divided and reflected.

A. Pons Varolii.
B. Medulla oblongata.
C. Pars cervicalis medullae spinalis.
D. Pars thoracica medullae spinalis.
E. Pars lumbalis medullae spinalis.
a. Crura cerebelli ad pontem.
b. Corpora pyramidalia.
c. Corpora olivaria.
d. Fissura mediana anterior.
e. Cervical expansion.
f. Lumbar expansion.
g. Bulbous or ganglionic expansion.
h. Filum terminale.
i. Ligamentum denticulatum.
k. Radix anterior nervi spinalis.
l. Radix posterior nervi spinalis.
m. Ganglion of posterior roots forming sensory portion of spinal nerve.
n. Cauda equina.
  1—8) Nervi cervicales (1—8).
  9—20) Nervi thoracici (12).
 21—25) Nervi lumbales (5).
 26—30) Nervi sacrales (5).
   31) Nervi coccygei.

Tab. XXIV

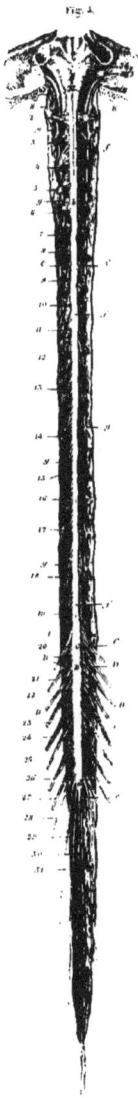

Fig. 4.

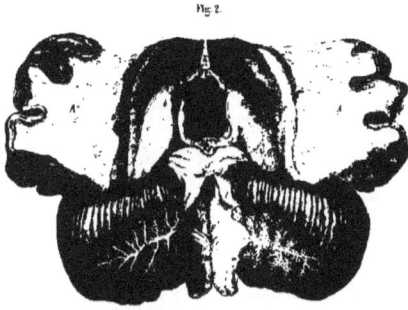

Fig. 2.

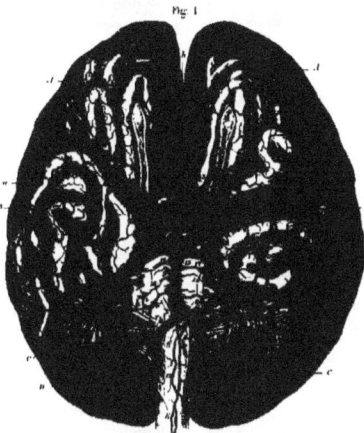

Fig. 1.

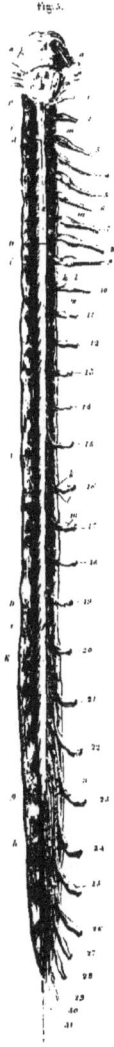

Fig. 5.

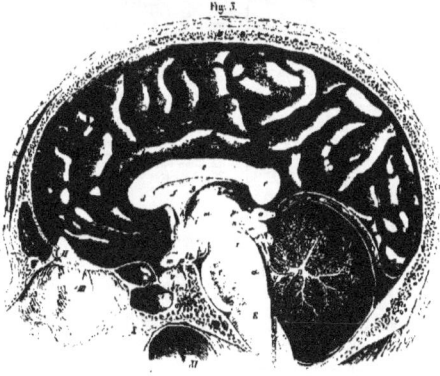

Fig. 3.

PLATE XXV.

NERVES OF THE HEAD AND NECK.

Fig. 1.

Internal view of Base or Cerebral Surface of Cranium, with Roots of Cerebral Nerves and exits from Cranium.

a Os frontis (pars frontalis).
b Sinus frontalis.
c Spina frontalis interna.
d Foramen caecum.
e Crista galli.
f Pars orbitalis ossis frontis.
g Cellulae pars ethmoidei.
h Corpus ossis sphenoidei.
i Ala magna ossis sphenoidei.
k Pars basilaris ossis occipitis.
l Pars squamosa ossis temporum.
m Pars petrosa ossis temporum.
n Foramen opticum.
o Foramen rotundum.
p Foramen ovale.
q Foramen spinosum.
r Fissura orbitalis superior.
s Cavitas tympani.
t Meatus auditorius internus.
u Malleolus (capitulum).
v Incus.
w Cochlea.
x Canalis semicircularis superior.
y Bulbus oculi.
z Glandula lacrymalis.
a M. rectus oculi externus.
β M. levator palpebrae superioris.
δ M. rectus oculi superior.
δ M. obliquus oculi superior.
δ* Trochlea pro m. obliquo superiore.
s M. rectus oculi internus.
ζ M. temporalis.
η M. pterygoideus externus.
θ Glandula pituitaria.
i Foramen magnum.
λ Foramen jugulare.
λ Foramen condyloideum anterius.
μ Pars occipitalis ossis occipitis (fossa cerebelli).
ν Sinus transversus.
   1. Art. meningea media.
   2. Art. carotis interna.
   3. Art. lacrymalis.
   4. Art. muscularis.
   5. Art. supraorbitalis.
   6. Art. ethmoidalis.
   7. Art ophthalmica.
   8. Nerv. opticus.
   9. Nerv. oculo-motorius (third nerve).
   10. Nerv. trochlearis (fourth nerve).
   11. Nerv. trigeminus (fifth nerve).
   12. Ganglion semilunare v. Gasseri.
   13. Nerv. maxillaris inferior, (3) — Branches of
   14. Nerv. maxillaris superior, (2) — fifth nerve
   15. Nerv. ophthalmicus, (1). — (nerv. trigeminus).
   16. Ramus frontalis — nerv.
   17. Ramus lacrymalis — ophthal-
   18. Ramus nasalis — mici.
   19. Nerv. ethmoidalis — nerv.
   20. Nerv. infratrochlearis. — nasalis.
   21. Nerv. supratrochlearis. } nerv.
   22. Nerv. supraorbitalis. } fronta- lis.
   23. Nerv. zygomaticus } nerv. la-
   24. Nerv. lacrymalis } crymalis.
   25. Nervi ciliares (from ganglion ophthalmicum).
   26. Nerv. buccinatorius.
   27. Nervi temporales profundi.
   28. Nerv. massetericus.
   29. Nerv. auricularis anterior v. temporalis superficialis.
   30. Art. maxillaris interna.
   31. Nn. meatus auditorii externi.
   32. Chorda tympani (nervi facialis).
   33. Nerv. petrosus superficialis minor.
   34. Nerv. Vidianus.
   35. Nerv. abducens (sixth nerve).
   36. Nerv. facialis (traversing meatus auditorius internus).
   37. Genu v. intumescentia gangliformis nervi facialis.
   38. Nerv. facialis in canalis Fallopii.
   39. Nerv. cochlae } nerv.
   40. Nerv. vestibuli } acusticae.
   41. Nerv. glossopharyngeus.
   42. Nerv. vagus.
   43. Nerv. accessorius Willisii.
   44. Nerv. hypoglossus.
   45. Arteria vertebralis.

Fig. 2.

Distribution of Superficial Nerves on the lateral surface of the Head and Neck.

a Os malare.
b Arcus zygomaticus.
c Maxilla inferior.
d Maxl. frontalis.
e M. orbicularis palpebrarum.
f M. zygomaticus major.
g M. zygomaticus minor.
h M. levator anguli oris.
i M. orbicularis oris.
k M. levator labii superioris.
l M. masseter.
m M. buccinator.
n M. triangularis menti v. depressor anguli oris.
o M. quadratus menti v. depressor labii inferioris.
p M. digastricus.
q Os hyoides.
r M. sterno-cleido-mastoideus.
s M. attollens auris.
t Mm. retrahentes auris.
u M. occipitalis.
v M. trapezius.
w M. splenius capitis.
x M. splenius colli.
y M. levator anguli scapulae.
z M. digastricus.
a Glandula submaxillaris.
β Ductus Stenonianus.
γ Galea aponeurotica.
δ Clavicula.
ε Acromion.
ζ Spina scapulae.
η Macl. deltoideus.
θ M. sternohyoideus.
i M. omohyoideus.
k M. scalenus anticus.
λ M. mylohyoideus.
   1. Arteria carotis communis.
   2. Art. carotis interna.
   3. Art. carotis externa.
   4. Art thyroidea superior.
   5. Art. maxillaris externa v. facialis.
   6. Art temporalis.
   7. Ramus frontalis } arteriae
   8. Ramus temporalis } tempora-
   9. Ramus occipitalis } lis.
   10. Art. transversalis faciei.
   11. Art. frontalis.
   12. Art. ophthalmica.
   13. Art. occipitalis.
   14. Nervus supraorbitalis.
   15. Nerv. supratrochlearis.
   16. Nerv. infratrochlearis.
   17. Nerv. infraorbitalis (with nn. palpebrales inferiores, subcutanei nasi et labiales superiores).
   18. Nerv. cutaneus malae.
   19. Nerv. mentalis (with nn. labiales inferiores).
   20. Nerv. ethmoidalis.
   21. Nerv. auricularis anterior v. temporalis superficialis.
   22. Nerv. facialis.
   23. Nerv. auricularis posterior.
   24. Ramus superior } nervi
   25. Ramus inferior } facialis.
   26. Nervi temporales faciales.
   27. Nervi zygomatici faciales.
   28. Nervi buccales v. rami faciales nervi facialis.
   29. Nerv. marginalis v. subcutaneus maxillae inferioris.
   30. Nerv. subcutaneus colli superior.
   31. Nerv. occipitalis major.
   32. Nerv. occipitalis minor.
   33. Nerv. auricularis superior.
   34. Nerv. auricularis magnus v. inferior.
   35. Nervi subcutanei colli inferiores.
   36. Nervi subcutanei colli medii.
   37. Nervi supraclaviculares.
   38. Nerv. accessorius Willisii.
   39. Plexus brachialis.

Fig. 3.

Distribution of Deep Nerves of Head, Neck, and Orbit.

a Os frontis.
b Skin of forehead and upper eyelids.
c Bulbus oculi.
d Membrana sinus maxillaris.
e Maxilla inferior.
f Processus pterygoideus.
g Ala magna ossis sphenoidei.
h Muscl. pterygoideus internus.
i Membrana tympani.
k Macl. stylohyoideus.
l M. digastricus.
m M. sternohyoideus.
n M. omohyoideus.
o M. sternothyroideus.
p M. hyothyroideus.
q M. mylohyoideus.
r M. sternocleidomastoideus.
s M. trapezius.
t M. splenius capitis.
u M. levator anguli scapulae.
v Glandula thyroidea.
   1. Vena jugularis interna.
   2. Arteria carotis communis.
   3. Art. thyroidea superior.
   4. Art. maxillaris externa facialis.
   5. Art. maxillaris interna.
   6. Art. alveolaris inferior.
   7. Art. alveolaris superior posterior.
   8. Art. meningea media.
   9. Art. infraorbitalis.
   10. Art. alveolaris anterior.
   11. Art. supraorbitalis et nervus supraorbitalis.
   12. Nervus opticus.
   13. Nerv. oculomotorius (third nerve).
   14. Nerv. trochlearis (fourth nerve).
   15. Nerv. trigeminus (fifth nerve).
   16. Ganglion Gasseri.
   17. Ramus (1) }
   18. Ramus (2) } nervi trigemini.
   19. Ramus (3) }
   20. Nerv. frontalis.
   21. Nerv. lacrymalis.
   22. Ramus internus } nerv.
   23. Ramus externus } lacrymalis.
   24. Nerv. subcutaneus malae.
   25. Nerv. ethmoidalis.
   26. Nerv. infraorbitalis.
   27. Nerv. alveolaris v. dentalis posterior.
   28. Nerv. alveolaris medius.
   29. Nerv. alveolaris anterior.
   30. Plexus dentalis superior.
   31. Nerv. lingualis (ninth nerve).
   32. Nerv. alveolaris inferior.
   33. Ganglion oticum et nerv. auricularis anterior.
   34. Connexion between otic ganglion and tympanic branches of the ganglion petrosum.
   35. Nerv. facialis.
   36. Intumescentia gangliformis nervi facialis.
   37. Chorda tympani.
   38. Nerv. Vidianus superficialis.
   39. Nerv. digastricus v. stylohyoideus.
   40. Nerv. hypoglossus (ninth nerve).
   41. Nerv. vagus.
   42. Nerv. laryngeus superior.
   43. Nerv. hypoglossus.
   44. Ramus descendens nervi hypoglossi.
   45. Nervi subcutanei colli.
   46. Nervi supraclaviculares (anteriores, medii et posteriores).
   47. Plexus cervicalis.
   48. Nerv. auricularis magnus.
   49. Nerv. occipitalis minor.
   50. Nerv. occipitalis major.
   51. Nerv. accessorius Willisii.

Fig. 4.

Distribution of Deepest Nerves of Head, Neck, and Orbit.

a Os frontis.
b Integuments of forehead.
c Glandula lacrymalis.
d Bulbus oculi.
e Os maxillare superius.
f Os maxillare inferius.
g Macl. rectus oculi superior.
h M. rectus oculi inferior.
i M. buccinator.
k Lingua (tongue).
l Membrana mucosa oris.
m Macl. pterygoideus externus.
n M. pterygoideus internus.
o M. hyoglossus.
p M. mylohyoideus.
q M. geniohyoideus.
r M. digastricus.
s Glandula submaxillaris.
t Glandula sublingualis.
u Ductus Whartonianus.
v Glandula thyroidea.
w Pharynx.
x Macl. sternocleidomastoideus.
y M. obliquus superior.
z M. obliquus inferior.
a M. rectus capitis posticus major.
β M. multifidus spinae.
γ Processus transversus atlantis.
δ Clavicula.
ε Macl. supraspinatus.
ζ First rib.
η M. rhomboideus minor.
θ M. rhomboideus major.
i Spina scapulae.
k Processus spinosus of dentata.
λ Processus spinosus vertebrae dorsi (1).
μ Larynx with macl. thyrohyoideus.
   1. Arteria carotis communis.
   2. Art. carotis externa.
   3. Art. carotis interna.
   4. Art. thyroidea.
   5. Art. lingualis.
   6. Art. maxillaris externa v. facialis.
   7. Art. submentalis.
   8. Art. occipitalis.
   9. Art. temporalis.
   10. Art. maxillaris interna.
   11. Art. alveolaris superior posterior.
   12. Nervus oculomotorius (third nerve).
   13. Ganglion ophthalmicum with
   14. Nervuli ciliares.
   15. Ganglion semilunare nervi trigemini (fifth nerve).
   16. Ramus ophthalmicus
   17. Ramus maxillaris superior } nervi trigemini
   18. Ramus maxillaris inferior } (fifth nerve).
   18.* Portio minor v. nerv. krotaphitico-buccinatorius.
   19. Nervus frontalis.
   20. Nerv. supratrochlearis (fourth nerve).
   21. Nerv. supraorbitalis.
   22. Nerv. subcutaneus malae.
   23. Nerv. infraorbitalis.
   24. Ramus buccalis nervi alveolaris posterior.
   25. Nerv. alveolaris posterior.
   26. Nerv. buccinatorius.
   27. Nerv. Vidianus superficialis.
   28. Ganglion oticum v. auriculare.
   29. Nerv. petrosus superficialis minor; nervi tensoris tympani.
   30. Nerv. lingualis v. gustatorius (of fifth nerve).
   31. Ganglion maxillare.
   32. Chorda tympani.
   33. Rami linguales nervi gustatorii.
   34. Genu nervi facialis.
   35. Nerv. facialis.
   36. Nerv. glossopharyngeus.
   37. Nerv. tympanicus v. Jacobsonii.
   38. Nerv. carotico-tympanicus.
   39. Nerv. vagus.
   40. Nerv. laryngeus superior.
   41. Plexus gangliformis.
   42. Nerv. accessorius Willisii.
   43. Nerv. hypoglossus (ninth nerve).
   44. Ramus geniohyoideus hypoglossi.
   45. Nerv. cervicalis (2).
   46. Nerv. cervicalis (3).
   47. Nerv. cervicalis (4).
   48. Nerv. cervicalis (5).
   49. Nerv. cervicalis (6).
   50. Nerv. cervicalis (7).
   51. Nerv. occipitalis magnus.
   52. Nerv. dorsalis scapulae.
   53. Nerv. suprascapularis.
   54. Rami communicantes nervorum spinalium.
   55. Nerv. sympathicus (ganglion cervicale primum).
   56. Plexus caroticus nervorum mollium (nerve molles).

Tab.XXV.

Fig. 3

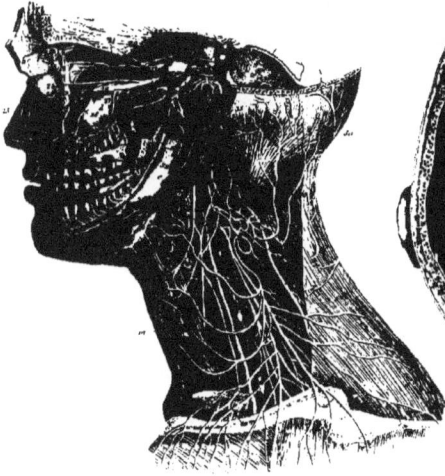

Fig 1.

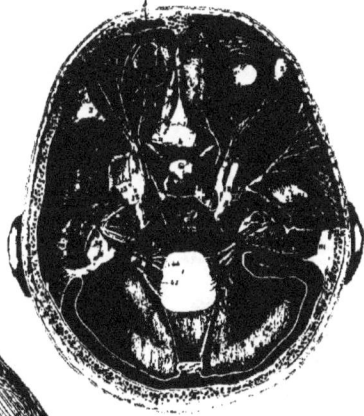

Fig. 2.

Fig. 4

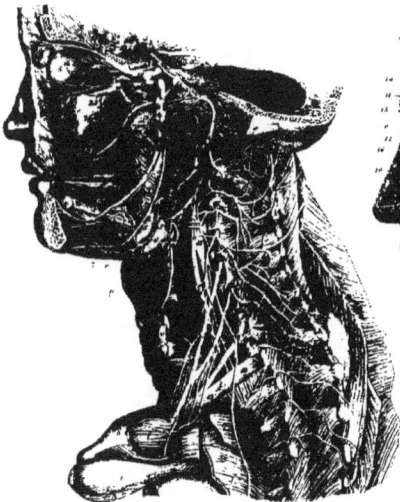

PLATE XXVI.

NERVES OF INTERIOR OF HEAD, WITH ANTERIOR VIEW OF MEDULLA SPINALIS AND SPINAL NERVES, SYMPATHETIC OR GANGLIONIC TRUNK, AND NERVES OF PELVIS.

Fig. 1.

Nerves of deep Muscles of Head, portions of the Zygoma and Ramus of Inferior Maxilla being removed (left side).

a Maxilla inferior.
b Temporo-maxillary articulation.
c Section of zygoma.
d Os malare.
e Os maxillare superius.
f Processus transversus atlantis.
g Processus styloideus.
h Meatus auditorius externus.
i Musculus temporalis.
k Mscl. masseter.
l M. pterygoideus externus.
m M. pterygoideus internus.
n M. buccinator.
o M. orbicularis oris.
p M. digastricus.
  1. Arteria carotis communis.
  2. Carotis externa.
  3. Carotis interna.
  4. Art. temporalis.
  5. Art. maxillaris externa v. facialis
  6. Art. maxillaris interna.
  7. Art. temporalis profunda.
  8. Nervus facialis.
  9. Ramus superior nervi facialis.
  10. Ramus inferior nervi facialis.
  11. Nerv. marginalis.
  12. Nerv. subcutaneus colli superior.
  13. Nerv. auricularis anterior.
  14. Nerv. buccinatorius.
  15. Nerv. massetericus.
  16. Nervi temporales profundi.
  17. Nerv. maxillaris inferior.
  18. Nerv. mentalis.

Fig. 2.

Deep branches of the three divisions of the Fifth Pair of Nerves (Trigemini), with the Nerves of the Tympanum.

a Ala magna ossis sphenoidei.
b Os malare.
c Os maxillare superius.
d Maxilla inferior.
e Pars petrosa ossis temporum.
f Processus mastoideus.
g Lingua (tongue).
h Glandula submaxillaris.
i Glandula sublingualis.
k Mscl. pterygoideus internus.
l M. pterygoideus externus.
m M. genioglossus.
n M. hyoglossus.
  1. Arteria carotis interna.
  2. Nervus trigeminus.
  3. Ganglion semilunare v. Gasseri.
  4. Ramus ophthalmicus nervi trigemini.
  5. Ramus maxillaris superior nervi trigemini.
  6. Ramus maxillaris inferior nervi trigemini.
  7. Nerv. facialis.
  8. Nerv. Vidianus superficialis.
  9. Nerv. maxillaris inferior.
  10. Nerv. mentalis.
  11. Nerv. mylohyoideus.
  12. Nerv. auriculares anterior (with art. meningea media).
  13. Nerv. gustatorius.
  14. Ganglion maxillare v. linguale.
  15. Nerv. temporalis profundus.
  16. Nerv. buccinatorius.
  17. Chorda tympani.
  18. Plexus caroticus internus (nervi sympathici), with nn. carotico-tympanici.

Fig. 3.

Vertical section of Bones of Face, showing distribution of Nasal Nerves and Branches of Third Division of Fifth Nerve (left side).

a Os frontis (sinus frontalis).
b Os nasale.
c Os ethmoideum (lamina cribrosa).
d Os sphenoideum (sinus sphenoidalis).
e Processus pterygoideus.
f Palatum durum.
g Canalis incisivus.
h Uvula.
i Septum nasi.
k Maxilla inferior.
l Foramen maxillare inferius.
m Musculus pterygoideus internus.
n Processus mastoideus.
  1. Arteria carotis interna (with plexus caroticus internus nervi sympathici).
  2. Nervus trigeminus.
  3. Ramus (1) nervi trigemini.
  4. Ramus (2) nervi trigemini.
  5. Ramus (3) nervi trigemini.
  6. Nervus nasopalatinus Scarpæ.
  7. Nerv. nasalis anterior (nerv. ethmoidal.).
  8. Nerv. lingualis (divided).
  9. Chorda tympani (divided).
  10. Nerv. maxillaris (with ramuli dentales).
  11. Nerv. mylohyoideus.

Fig. 4.

Vertical section of Bones of Face, showing internal distribution of the three divisions of Fifth Nerve (Trigemini).

a Os frontis (sinus frontalis).
b Pars orbitalis ossis frontis.
c Bulbus oculi.
d Mscl. levator palpebræ superioris.
e M. rectus oculi superior.
f M. rectus inferior.
g Nasal cavity and spongy bones.
h Processus pterygoideus.
i Foramen sphenopalatinum.
k Palatum durum.
l Canalis pterygopalatinus
m Maxilla inferior.
n Mscl. pterygoideus internus.
o Membrana tympani.
  1. Arteria maxillaris interna.
  2. Nervus opticus.
  3. Nerv. trigeminus with ganglion Gasseri.
  4. Ramus (1) nervi trigemini.
  5. Ramus (2) nervi trigemini.
  6. Ramus (3) nervi trigemini.
  7. Nerv. frontalis.
  8. Nerv. nasalis.
  9. Nerv. ethmoidalis.
  10. Ganglion ciliare v. lenticulare.
  11. Radix brevis ganglii ciliaris.
  12. Radix longa ganglii ciliaris.
  13. Nervuli ciliares v. ganglio ciliari.
  14. Nerv. sphenopalatinus.
  15. Ganglion sphenopalatinum, or Meckel's ganglion.
  16. Nervi nasales posteriores superiores.
  17. Nerv. pterygopalatinus.
  18. Nerv. nasalis posteriores inferiores.
  19. Nerv. Vidianus superficialis.
  20. Nerv. gustatorius.
  21. Ganglion maxillare v. linguale.
  22. Nerv. pterygoideus.
  23. Nerv. facialis (in canalis Fallopii).
  24. Chorda tympani.
  25. Nerv. auricularis anterior.
  26. Ganglion oticum.

Fig. 5.

Anterior view of Pons Varolii, Medulla Oblongata, Medulla Spinalis and Spinal Nerves, Sympathetic Ganglia, and Pelvic Nerves.

a Pars petrosa ossis temporum.
b Meatus auditorius internus.
c Canalis caroticus.
d Processus styloideus.
e Foramen jugulare.
f Meatus auditorius externus.
g Atlas.
h Vertebra cervici (7).
i Vertebra dorsal (1).
k Vertebra dorsi (12).
l Vertebra lumbalis (1).
m Vertebra lumbalis (3).
n Os sacrum.
o Os coccygis.
p First rib.
q Last rib.
r Crista ossis ilii.
s M. sterno-cleido-mastoideus.
t M. scalenus anticus.
u M. scalenus medius.
v Mm. intercostales interni.
w Mm. intercostales externi.
x M. quadratus lumborum.
y M. psoas major.
z M. iliacus internus.
  1. Art. carotis interna.
  2. Vena jugularis interna.
  3. Art. v. ven. intercostalis (posterior).
  4. Pons Varolii.
  5. Nerv. trigeminus (divided).
  6. Nerv. abducens.
  7. Nerv. facialis et acousticus.
  8. Nerv. glossopharyngeus, vagus et accessorius (passing out through the internal jugular foramen).
  9. Nerv. accessorius Willisii.
  10. Nerv. vagus v. pneumo-gastricus.
  11. Nerv. hypoglossus v. lingualis.
  12. Ramus descendens nervi hypoglossi.
  13. Medulla oblongata.
  14. Decussatio pyramidum.
  15. Pars cervicalis medullæ spinalis.
  16. Pars thoracica medullæ spinalis.
  17. Bulbous expansion at end of spinal cord.
  18. Filum terminale.
  19. Nerv. cervicalis (1).
  20. Nerv. cervicalis (8).
  21. Plexus cervicalis.
  22. Plexus brachialis.
  23. Nerv. dorsalis (1).
  24. Nerv. dorsalis (12).
  25. Nn. intercostales.
  26. Nerv. lumbalis (1).
  27. Nerv. lumbalis (5).
  28. Plexus lumbalis.
  29. Nerv. cruralis anterior.
  30. Nerv. ilio-hypogastricus (ramus externus et internus).
  31. Nerv. ilio-inguinalis.
  32. Nerv. inguino-cutaneus.
  33. Nerv. obturatorius.
  34. Nerv. sacralis (1).
  35. Nerv. sacralis (5).
  36. Plexus sacralis.
  37. Nn. coccygei.
  38. Nerv. sympathicus.
  39. Ganglion cervicale supremum.
  40. Ganglion cervicale medium.
  41. Ganglion cervicale infimum.
  42. Ganglia thoracica.
  43. Ganglia lumbalia.
  44. Ganglia sacralia.
  45. Ganglion coccygeum (or ganglion impar).
  46. Connecting branches between sacral sympathetic nerves of right and left sides.
  47. Nervus ischiadicus.
  48. Nerv. inguinalis.

Tab. XXVI.

Fig. 5.

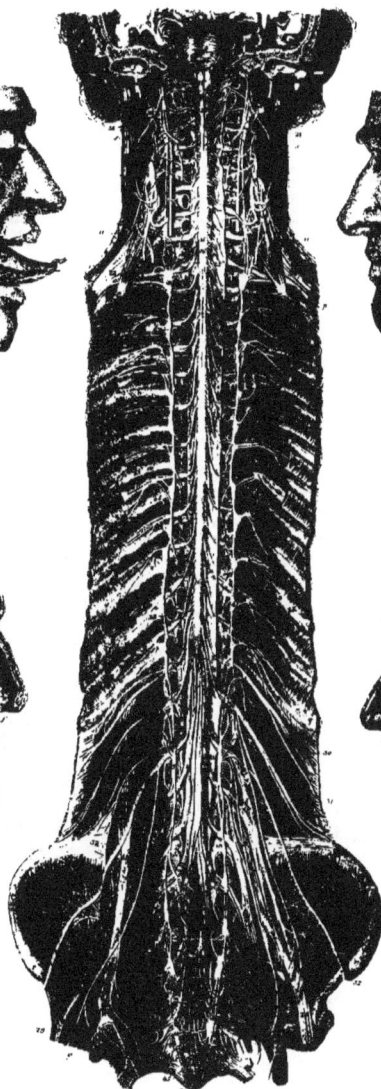

Fig. 2.

Fig. 1.

Fig. 3.

Fig. 4.

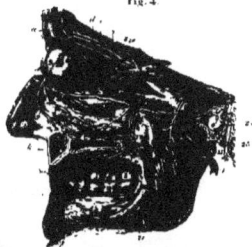

A. Vonder zu Leipzig

PLATE XXVII.

NERVES OF THE POSTERIOR AND LATERAL SURFACE OF TRUNK, DEEP PELVIC, AND SUPERFICIAL ANO-PERINEAL REGIONS.

Fig. 1.

Superficial and deep Nerves on dorsum of Trunk.

a   Processus mastoideus.
b   Spina scapulæ.
c   Crista ossis ilii.
d   Macl. occipitalis.
e   M. trapezius.
f   M. deltoideus.
g   M. latissimus dorsi.
h   M. infraspinatus.
i   M. obliquus abdominis externus.
k   M. sacro-lumbalis.
l   M. lumbo-costalis.
m   M. longissimus dorsi.
n   M. cervicalis ascendens.
o   M. transversalis cervicis.
p   M. trachelo-mastoideus.
q   M. complexus cervicis.
r   M. biventer cervicis.
s   M. trapezius.
t   M. obliquus capitis superior.
u   M. obliquus capitis inferior.
v   M. levator costæ.
w   M. transversalis abdominis.

1. Nervus occipitalis magnus.
2. Nerv. occipitalis minor.
3. Nerv. auricularis superior.
4. Rami posteriores nervorum cervicalium.
5. Rami posteriores nervorum dorsalium with rami externi and interni.
6. Nervi supraclaviculares posteriores.
7. Nerv. cutaneus brachii posterior superior.
8. Ramus posterior nerv. cutanei brachii intern, posterior (of the nerv. intercostal (2).
9. Rami posteriores nervorum cutan. pectoris (from posterior nerv. intercostal).
10. Rami posteriores nervor. cutan. abdominis (from posterior nervor. intercostal).
11. Rami externi dorsales (from poster. and inferior nerv. dorsal.).
12. Rami posteriores nervor. lumbalium.
13. Nervi glutæi superiores posteriores.

Fig. 2.

Nerves of Right Axillary Cavity and lateral surface of Trunk.

a   Clavicula.
b   Macl. scalenus anticus.
c   M. scalenus medius.
d   M. levator anguli scapulæ.
e   M. deltoideus.
f   M. pectoralis major.
g   M. pectoralis minor.
h   Processus coracoideus.

i   M. latissimus dorsi.
k   M. biceps (caput breve).
l   M. coraco-brachialis.
m   M. serratus anticus major.
n   M. subscapularis.
o   M. serratus posticus inferior.
p   Mm. intercostales.
q   M. transversalis abdominis.
r   M. rectus abdominis.
s   M. obliquus externus abdominis.
t   M. obliquus internus abdominis.
u   Art. carotis communis.
v   Art. subclavia.
w   Art. axillaris.
x   Vena jugularis interna.
y   Ven. subclavia.
z   Ven. axillaris.
a   Art. et ven. brachialis.
β   Art. et ven. subscapularis.
γ   Art. et ven. thoracica longa.
δ   Art. et ven. thoracico-dorsalis.

1. Nerv. cervicalis (5).
2. Nerv. cervicalis (6).
3. Nerv. cervicalis (7).   Anterior branches forming cervical plexus.
4. Nerv. cervicalis (8).
5. Nerv. dorsalis (1).
6. Nervus phrenicus.
7. Ramus nerv. cervicalis (4) ad m. levatorem scapulæ.
8. Nerv. suprascapularis v. scapularis.
9. Nervi thoracici externi.
10. Nerv. thoracicus longus.
11. Ramus anterior v. thoracicus nerv. pectoralis externi (from second intercostal nerve).
12. Nerv. cutaneus internus brachii posterior (from second intercostal nerve—nerves of Wrisberg).
13. Nervi subscapulares.
14. Nerv. cutaneus brachii internus.
15. Nerv. cutaneus brachii medius.
16. Nervus cutaneus brachii externus v. musculo-cutaneus v. perforans Casseri (perforating m. coraco-brachialis).
17. Nerv. axillaris v. circumflexus humeri.
18. Heads of nerv. medianus.
19. Nerv. medianus.
20. Nerv. ulnaris.
21. Nerv. radialis.
22. Nn. pectorales externi from 2—7 nerv. intercost.
23. Nn. pectorales interni from 2—7 nerv. intercost.
24. Nn. abdominales interni.

Fig. 3.

Superficial Nerves of Ano-Perineal Region and Scrotum.

a   Tuber ischii.
b   Ramus ascendens ossis ischii.

c   Great sacro-sciatic ligament.
d   Corpus spongiosum urethræ.
e   Corpus cavernosum penis.
f   Anus.
g   Scrotum.
h   Macl. acceleratores urinæ.
i   M. ischio-cavernosus (erector penis).
k   M. transversus perinæi.
l   M. sphincter ani externus.
m   M. levator ani.
n   M. adductor longus.
o   M. Gracilis.
p   M. adductor magnus.
q   M. Glutæus maximus.

1. Nerv. pudendus communis.
2. Nerv. perinæi v. pudendus internus.
3. Nerv. dorsalis penis.
4. Nn. scrotales inferiores.
5. Nn. glutæi inferiores.
6. Nn. subcutanei perinæi.

Fig. 4.

Deep Pelvic Nerves and branches to side of Bladder and Penis—left surface.

a   Vertebra lumbalis.
b   Os sacrum.
c   Os coccygis.
d   Intestinum rectum.
e   Macl. sphincter ani internus.
f   M. levator ani.
g   M. erector penis.
h   Mm. acceleratores urinæ.
i   Corpus spongiosum urethræ.
k   Corpus cavernosum penis.
l   Glans penis.
m   Scrotum.
n   Symphysis ossium pubis.
o   Mons Veneris.
p   Vesica urinaria.
q   Peritonæal surface of bladder.
r   Pars membranacea urethræ.
s   Glandula prostata.
t   Ureter.
u   Vas deferens.
v   Peritonæum.
w   Excavatio recto-vesicalis.
x   Arteria iliaca sinistra (with plexus hypogastricus).
y   Art. iliaca externa.
z   Art. iliaca interna.
a   Art. vesicalis.
1. Plexus pudendo-hæmorrhoidalis.
2. Rami anteriores nervor. sacral.
3. Nervi hæmorrhoidales (medii et inferiores).
4. Nervi vesicales.
5. Nerv. dorsalis penis.
6. Pars lumbalis nervi sympathici.

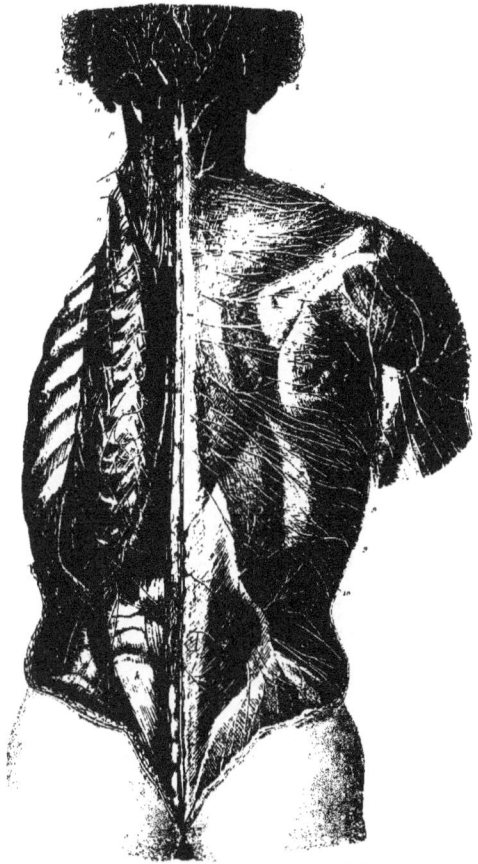

Fig. 1.

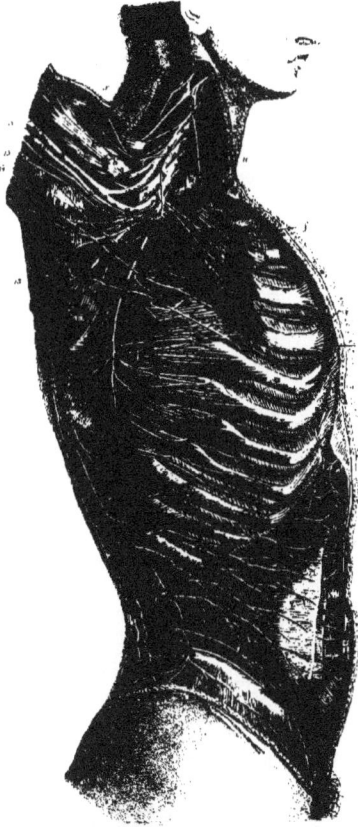

Fig. 2.

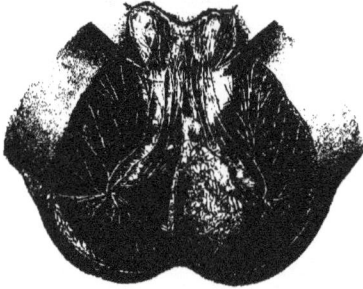

Fig. 3.

Fig. 4.

PLATE XXVIII.

SUPERFICIAL AND DEEP NERVES OF POSTERIOR SURFACE OF SHOULDER AND UPPER EXTREMITIES, AND OF ANTERIOR SURFACE OF LOWER EXTREMITIES.

Fig. 1.

Superficial Nerves on posterior surface of Upper Extremity—Right.

a Mscl. deltoideus.
b M. latissimus dorsi.
c M. triceps humeri.
d Olecranon.
e Condylus externus ossis brachii.
f Capitulum ulnæ.
g Vena cephalica brachii.
h Vena cephalica pollicis.
i Vena basilica.
1. Nervi supraclaviculares posteriores.
2. Nerv. cutaneus brachii posterior superior.
3. Cutaneous branches of axillary plexus.
4. Nerv. cutaneus antibrachii externus superior.
5. Nerv. cutaneus brachii internus posterior.
6. Nerv. cutaneus brachii internus minor.
7. Ramus cutaneus (externus) nervi musculo-cutanei.
8. Ramus v. nerv. dorsalis radialis, with rami digitales dorsales for first, second, and third finger.
9. Ramus posterior v. ulnaris of nerv. cutaneus brachii medius.
10. Ramus dorsalis nervi ulnaris (with rami digitales dorsales for fourth and fifth finger).

Fig. 2.

Deep Nerves on posterior surface of Upper Extremity—Left.

a Acromion scapulæ.
b Spina scapulæ.
c Caput ossis humeri.
d Olecranon.
e Ulna.
f Radius.
g Os humeri.
h Ligamentum transversum v. scapulæ proprium posticum.
i Musculus supraspinatus.
k Mscl. infraspinatus.
l M. teres major.

m M. teres minor.
n M. deltoideus.
o M. triceps humeri.
p M. biceps and brachialis internus.
q Condylus internus humeri.
r M. supinator longus.
s Mm. extensores carpi radiales.
t M. supinator brevis.
u Mm. flexores carpi.
v M. extensor pollicis longus.
w Mm. extensores primi and secundi internodii pollicis.
x Tendo extensor digitorum communis.
y Posterior annular ligament.
1. Art. and ven. profunda brachii.
2. Art. radialis dorsalis.
3. Ven. cephalica brachii.
4. Ven. basilica.
5. Nervus suprascapularis.
6. Nerv. supraspinatus of nervus suprascapularis.
7. Nerv. infraspinatus of nervus suprascapularis.
8. Nerv. axillaris.
9. Nerv. cutaneus brachii posterior superior.
10. Nerv. musculo-spiralis v. radialis.
11. Nerv. ulnaris.
12. Nerv. interosseous externus v. ramus profundus nervi radialis.
13. Nerv. radialis superficialis.
14. Nerv. cutaneus antibrachii externus superior.
15. Ramus dorsalis nervi ulnaris.

Fig. 3.

Superficial and Deep Nerves on anterior surface of Lower Extremities.

a Promontorium ossis sacri.
b Crista ilii.
c Spina ilii anterior superior.
d Symphysis ossium pubis.
e Mscl. iliacus internus.
f M. psoas major.
g M. sartorius.
h M. gluteus medius.
i M. tensor fasciæ latæ.
k M. rectus femoris.
l M. vastus externus.

m M. pectinæus.
n M. adductor longus.
o M. adductor magnus.
p M. cruralis.
q M. vastus internus.
r Tendo extensorius cruris communis.
s Patella.
t Tibia.
u Malleolus internus.
v Malleolus externus.
w Ligamentum transversum v. vaginale cruris.
x M. tibialis anticus.
y M. extensor digitorum pedis longus.
z M. peronæus longus.
α M. peronæus brevis.
β M. extensor pollicis pedis longus.
γ M. extensor digitorum pedis brevis.
δ M. extensor pollicis pedis brevis.
ε M. soleus.
1. Aorta abdominalis.
2. Arteria iliaca (communis).
3. Art. iliaca interna.
4. Art. femoralis.
5. Vena femoralis.
6. Ven. saphena magna.
7. Ven. saphena parva.
8. Plexus ischiadicus.
9. Plexus lumbalis.
10. Nervus cruralis anterior.
11. Nerv. obturatorius.
12. Nerv. cutaneus femoris anterior externus.
13. Nerv. inguinalis.
14. Nerv. inguino-cutaneus.
15. Nerv. saphenus major.
16. Nerv. cutaneus femoris anterior medius.
17. Nerv. cutaneus femoris anterior internus v. saphenus minor.
18. Rami nervi ilio-hypogastrici.
19. Rami nervi ilio-inguinalis.
20. Rami musculares nervi cruralis.
21. Nerv. peronæus superficialis with
22. Nerv. cutaneus dorsi pedis internus, et
23. Nerv. cutaneus dorsi pedis medius.
24. Nerv. cutaneus cruris externus.
25. Nerv. peronæus profundus.
26. Ramus internus nervus peronæus profundus.
27. Ramus externus nervus peronæus profundus.
28. Pars sacralis nerv. sympathici.

Tab. XXVIII.

Fig. 1

Fig. 3

Fig. 2

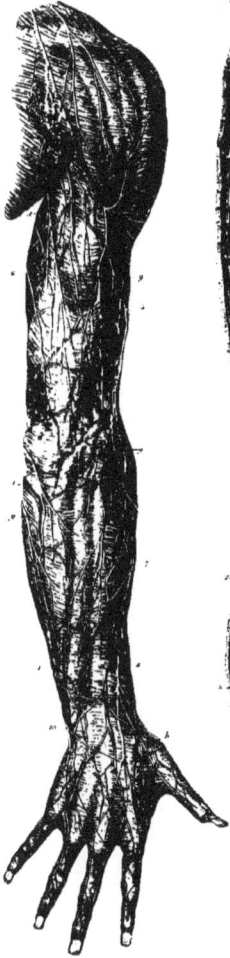

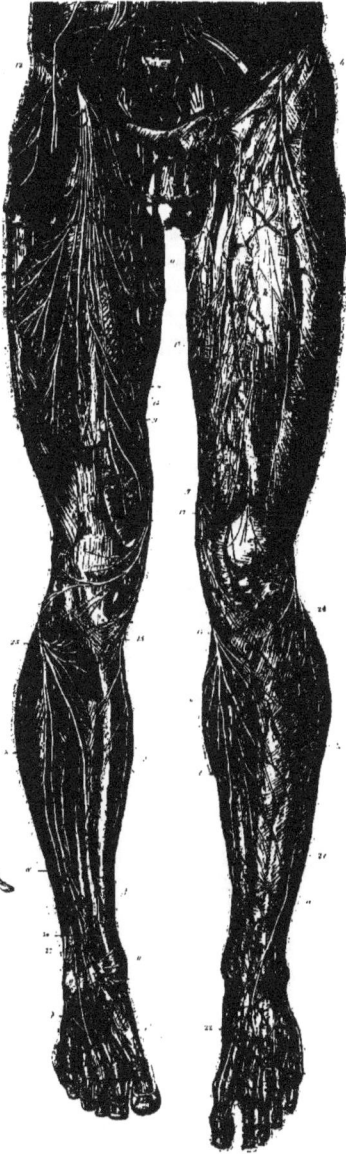

PLATE XXIX.

SUPERFICIAL AND DEEP NERVES ON THE ANTERIOR SURFACE OF UPPER EXTREMITIES, AND POSTERIOR SURFACE OF THE LOWER EXTREMITIES.

Fig. 1.

Superficial Nerves on Anterior Surface of Upper Extremity—Right.

a   Macl. deltoidens.
b   M. pectoralis major.
c   M. biceps flexor cubiti.
d   Plica cubiti.
e   Capitulum ulnæ.
f   Aponeurosis palmaris.
g   Fleshy ball of thumb.
h   Fleshy ball of little finger and palmaris brevis m.
i   Vena cephalica brachii.
k   Ven. basilica.
l   Ven. mediana basilica.
m   Ven. mediana cephalica.
  1. Nervi supraclaviculares.
  2. Ramus cutaneus nervi axillaris.
  3. Nerv. cutaneus brachii internus posterior.
  4. Nerv. cutaneus brachii internus (minor).
  5. Ramus nervi cutanei medii.
  6. Nerv. cutaneus brachii medius v. internus major.
  7. Ramus cutaneus palmaris, nerv. cutan. medii.
  8. Ramus cutaneus ulnaris, nerv. cutan. medii.
  9. Ramus cutaneus nerv. musculo-cutanei.
  10. Ramus nervi radialis.
  11. Nerv. ulnaris volaris.
  12. Nervi digitales volares.

Fig. 2.

Deep Nerves on Anterior Surface of Upper Extremity—Left.

a   Caput ossis humeri.
b   Processus coracoideus.
c   Macl. deltoideus.
d   M. pectoralis major.
e   M. pectoralis minor.
f   M. biceps flexor cubiti.
g   Caput breve, macl. bicipitis.
h   Caput longum, macl. bicipitis.
i   M. coraco-brachialis.
k   M. brachialis internus.
l   Caput internum, musculi tricipitis extensoris.
m   Caput longum, musculi tricipitis extensoris.
n   M. supinator longus.
o   M. extensor carpi radialis longus.
p   M. pronator teres.
q   M. flexor carpi radialis.
r   M. supinator brevis.
s   Mm. flexores digitorum communes.
t   M. flexor carpi ulnaris.
u   M. flexor pollicis longus.
v   M. abductor and flexor brevis pollicis.
w   M. adductor pollicis.
x   M. abductor digiti minimi.
y   Ligamentum carpi volare profundum.
z   M. flexor digiti minimi.
  1. Arteria axillaris.
  2. Art. and ven. brachialis.
  3. Art. and ven. ulnaris.
  4. Art. and ven. radialis.
  5. Arcus palmaris sublimis.
  6. Plexus axillaris (brachialis).
  7. Nervus medianus (sending branches to both sides of thumb, index and middle fingers, and radial side of ring finger).
  8. Nerv. ulnaris.
  9. Nerv. ulnaris volaris (sending branches to ulnar side of ring, and both sides of little finger).
  10. Nerv. musculo-spiralis v. radialis.
  11. Nerv. interosseus externus.
  12. Nerv. radialis superficialis.
  13. Nerv. musculo-cutaneus.

Fig. 3.

Superficial and deep Nerves on Posterior Surface of Lower Extremities.

a   Muscl. glutæus maximus.
b   M. glutæus medius.
c   M. glutæus minimus.
d   M. pyriformis.
e   Mm. gemelli with obturator internus tendon.
f   M. quadratus femoris.
g   Tuber ischii.
h   Great sacro-sciatic ligament.
i   M. levator ani.
k   M. vastus externus.
l   M. biceps femoris.
m   Caput breve, macl. bicipitis.
n   Caput longum, macl. bicipitis.
o   M. semitendinosus.
p   M. semimembranosus.
q   M. gracilis.
r   M. sartorius.
s   M. adductor magnus.
t   M. gastrocnemius.
u   M. popliteus.
v   M. soleus.
w   Mm. peronæus longus and brevis.
x   M. flexor pollicis pedis longus.
y   Mm. tibialis posticus and flexor digitorum communis longus.
z   Tendo Achillis.
  1. Vena saphena parva.
  2. Art. and ven. tibialis postica.
  3. Nervi cutanei glutæi superiores (from posterior branches of lumbar nerves).
  4. Nerv. cutaneus femoris posterior.
  5. Nervi cutanei glutæi inferiores.
  6. Rami nervi sapheni minoris.
  7. Nerv. saphenus major.
  8. Nerv. communicans fibularis (v. saphenus externus).
  9. Nervi communicans tibialis.
  10. Nerv. cutaneus externus dorsi pedis.
  11. Nerv. glutæus superior.
  12. Nerv. glutæus inferior.
  13. Nerv. pudendus communis.
  14. Plexus ischiadicus.
  15. Nerv. cutaneus femoris posterior communis.
  16. Nerv. ischiadicus.
  17. Nerv. peronæus v. fibularis.
  18. Nerv. tibialis.
  19. Nerv. communicans fibularis.

Fig. 4.

Deep Nerves in Sole of Foot—Left.

a   Tuberositas calcanei.
b   Macl. abductor digiti minimi.
c   M. abductor pollicis pedis.
d   Tendo m. flexoris pollicis pedis longi.
e   M. flexor digitorum communis brevis.
f   Tendo Achillis.
g   Malleolus internus.
h   Malleolus externus.
  1. Nerv. tibialis.
  2. Nervi cutanei plantares.
  3. Nerv. plantaris internus.
  4. Nerv. plantaris externus.

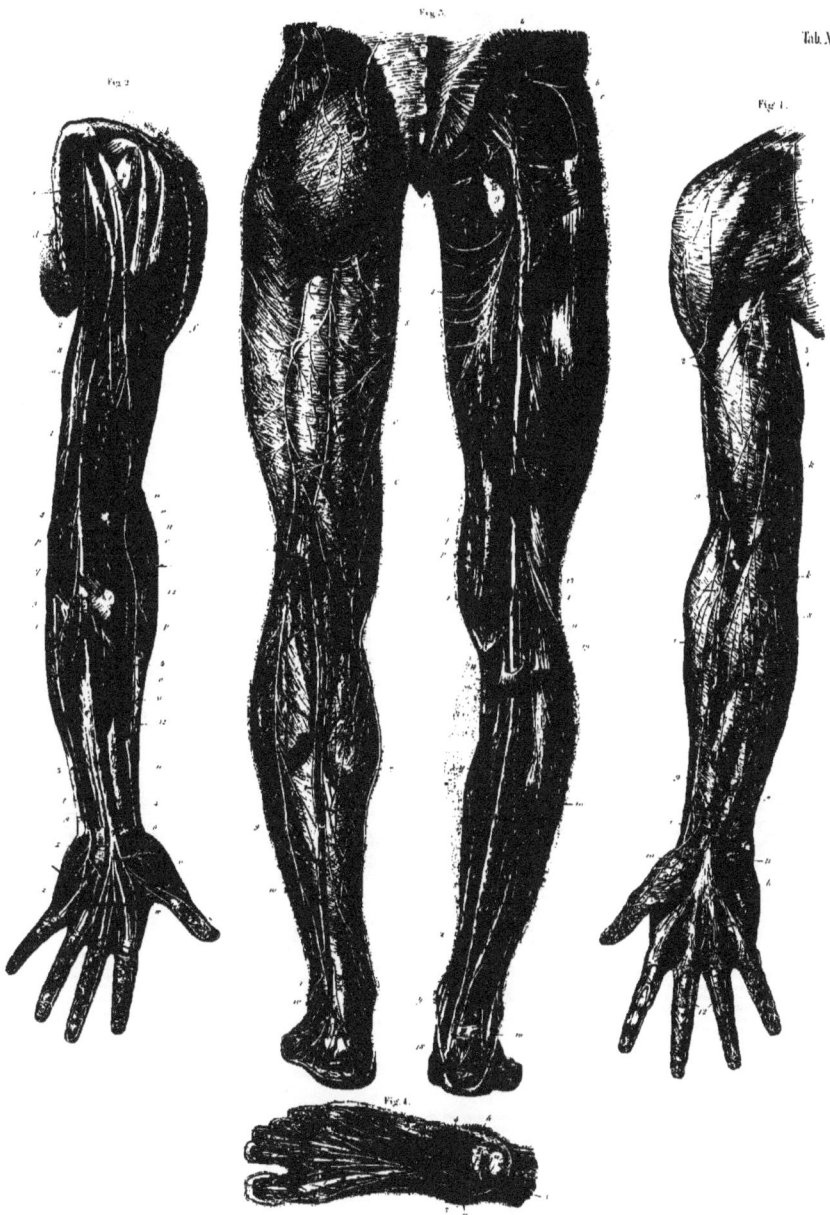

Tab. XXIX.

Fig. 3.

Fig. 2

Fig. 1.

Fig. 4.

A Anatomy Legacy

PLATE XXX.

SYMPATHETIC SYSTEM OF NERVES,
WITH THE PRINCIPAL CEREBRO-SPINAL NERVES IN
CONNECTION THEREWITH.

Fig. 1.

Nerves of left Orbit.

a Os frontis (with sinus frontalis).
b Os maxillare superius (with sinus maxillaris).
c Os sphenoidale.
d Bulbus oculi.
e Musculus rectus oculi externus.
f Mscl. rectus oculi inferior.
g M. obliquus oculi superior.
h M. obliquus inferior.
i M. rectus oculi superior.
k M. levator palpebræ superioris.
l Arteria carotis interna.
　1. Nervus opticus.
　2. Nerv. oculomotorius (third nerve).
　4. Ramus superior } nervi oculo-
　4. Ramus inferior } motorii.
　　With radix brevis ganglii ciliaris.
　5. Nerv. trochlearis (fourth nerve).
　6. Nerv. trigeminus (fifth nerve).
　7. Ganglion semilunare.
　8. Ramus (1), nervi trigemini.
　6. Ramus (2), nervi trigemini.
　10. Ramus (3), nervi trigemini.
　11. Nerv. frontalis.
　12. Nerv. nasociliaris with
　13. Radix longa of the
　14. Ganglion ciliare v. lenticulare.
　15. Nervoli ciliares.
　16. Nerv. ethmoidalis.
　17. Nerv. trochlearis.
　18. Nerv. abducens.

Fig. 2.

Vertical section of Head behind the left Ear, through the cavity of the Tympanum and Internal Ear, showing the Nerves in their course through these.

a Os parietale.
b Pars squamosa ossis temporalis.
c Pars mastoidea ossis temporalis.
d Pars petrosa ossis temporalis.
e Ala parva ossis sphenoidei.
f Corpus ossis sphenoidei.
g Processus pterygoideus.
h Maxilla inferior.
i Musculus pterygoideus internus.
k Mscl. tensor tympani.
l Membrana tympani (inner surface) with malleus and incus.
m Auricula sinistra.
n Parotis.
o Arteria carotis externa.
p Art. auricularis posterior.
q Art. stylomastoidea.
r Art. maxillaris interna.
s Art. auricularis profunda.
t Art. dentalis inferior.
u Art. meningea media.
v Art. pterygoidea.
　1. Nerv. trigeminus (fifth nerve).
　2. Portio major, nervi trigemini.
　3. Portio minor, nervi trigemini.
　4. Ganglion Gasseri v. semilunare.
　5. Ramus (1), nervi trigemini.
　6. Ramus (2), nervi trigemini.
　7. Ramus (3), nervi trigemini.
　8. Nerv. auricularis anterior.
　9. Nerv. dentalis inferior.
　10. Nerv. gustatorius.
　11. Nerv. pterygoideus.
　12. Ganglion oticum.
　13. Nervoli pro mscl. tensore palati mollis.
　14. Nervoli pro mscl. tensore tympani.
　15. Nerv. facialis (in canalis Fallopii).
　16. Chorda tympani.

Fig. 3.

Vertical section through Mastoid and Petrous portions of Temporal Bone, exposing cavity of Tympanum and Nerves traversing it.

a Processus mastoideus.
b Pars petrosa ossis temporum.
c Canalis Fallopii.
d Cavitas tympani.
e Incus.
f Malleus.
g Musculus tensor tympani.
　1. Nervus facialis (portio dura of seventh nerve).
　2. Genu nervi facialis.
　3. Chorda tympani.
　4. Nerv. trigeminus.
　5. Ganglion semilunare.
　6. Ramus (1), nervi trigemini.
　7. Ramus (2), nervi trigemini.
　8. Ramus (3), nervi trigemini.
　9. Nerv. auricularis anterior (with its double root).
　10. Ganglion oticum.
　11. Nervulus pro mscl. tensore tympani
　12. Nerv. petrosus superficialis minor.
　13. Nerv. petrosus superficialis major.

Fig. 4.

Cavity of left Tympanum, with course and branches of Facial Nerve through Fallopian Canal.

a Pars petrosa, ossis temporum.
b Pars mastoidea, ossis temporum.
c Promontorium.
d Fenestra ovalis.
e Fenestra rotunda.
f Tuba Eustachii.
g Musculus tensor tympani.
h Mscl. stapedius.
i Arteria carotis interna.
k Vena jugularis interna.
　1. Nervus facialis.
　2. Genu nervi facialis.
　3. Nerv. petrosus superficialis major.
　4. Nerv. petrosus superficialis minor.
　5. Nerv. glosso-pharyngeus.
　6. Nerv. tympanicus v. Jacobsonii, with branches to fenestra rotunda, ovalis, cochlea, and int. carotid plexus.
　7. Nerv. petrosus profunda minor.
　8. Ramus auricularis nervi vagi.

Fig. 5.

Nerves of the Palate and Tongue.

a Labium superius.
b Palatum durum.
c Velum palatinum v. palatum molle.
d Uvula.
e Arcus glosso-palatinus.
f Arcus pharyngo-palatinus.
g Tonsilla.
h Isthmus faucium.
i Radix linguæ.
　1. Nervi palatini (anterior, internus and externus).
　2. Nervus glosso-pharyngeus.
　3. Nervus gustatorius nervi trigemini.
　4. Ramus nervi glosso-pharyngei (pro mscl. glosso-palatino).

Fig. 6.

Posterior view of the distribution of the Par Vagum and Sympathetic Nerves, the Spinal Column being removed.

a Pars basilaris ossis occipitis.
b Sella turcica (processus clinoidei posteriores).
c Pars petrosa ossis temporum.
d Canalis caroticus.
e Processus mastoideus.
f First rib.
g Second rib.
h Mscl. sternocleido-mastoideus.
i, k, and l Pharynx (mscl. constrictor pharygis superior, medius and inferior).
m Œsophagus.
n Bronchial tubes.
o Arcus aortæ.
p Arteria carotis communis sinistra.
q Art. subclavia sinistra.
r Art. innominata.
s Art. carotis communis dextra.
t Art. subclavia dextra.
u Art. mammaria interna.
v Art. thyreoidea superior.
w Art. thyreoidea inferior.
x Art. pharyngea ascendens.
y Art. cervicalis ascendens.
z Art. carotis interna.
　1. Nerv. sympathicus.
　2. Ganglion cervicale supremum (with nerv. cardiacus longus, nervi laryngo-pharyngei, and connections with plexus cervicalis).
　3. Ganglion cervicale medium (with nerv. cardiacus medius and connections with plexus brachialis, nerv. vagus, recurrens and phrenicus).
　4. Ganglion cervicale infimum (with nerv. cardiacus magnus and connections with plexus brachialis, subclavius, and pulmonalis.
　5. Nerv. cervicalis (1).
　6. Nerv. cervicalis (2).
　7. Nerv. cervicalis (3).
　8. Nerv. cervicalis (4).
　9. Plexus brachialis.
　10. Nerv. intercostalis.
　11. Nerv. phrenicus.
　12. Nerv. vagus.
　13. Nerv. recurrens v. laryngeus inferior.
　14. Plexus pharyngeus (superior and inferior).
　15. Plexus œsophageus.
　16. Plexus pulmonalis.

Fig. 7.

Distribution of Facial Nerve (portio dura) and Par Vagum or Pneumogastric, with the Sympathetic Ganglia and Nerves on lateral surface of Head and Trunk.

a Os maxillare inferius.
b Os hyoides.
c Clavicula.
d First rib.
e Second rib.
f Eleventh rib.
g Processus transversus vertebræ lumbalis.
h Os sacrum.
i Os pubis (symphysis).
k Musculus zygomaticus major.
l Mscl. digastricus maxillæ inferioris.
m M. masseter.
n Glandula parotis.
o Glandula submaxillaris.
p M. sterno-hyoideus.
q M. scalenus anticus.
r M. scalenus medius and portions.
s Diaphragma.
t M. quadratus lumborum.
u Bronchus dexter.
v Kidney.
w Glandula suprarenalis.
x Œsophagus.
y Stomach.
z Intestinum jejunum.
α Intestinum colon.
β Intestinum rectum.
γ Vesica urinaria.
δ Ureter.
ε Prostate gland.
ζ Vas deferens.
η Spermatic cord.
θ Art. and ven. spermatica interna with plexus spermaticus internus.
κ Penis.
λ Scrotum and testiculus.
μ Vena azygos.
ν Vena cava superior.
ρ Ductus thoracicus.
　1. Aorta descendens thoracica.
　2. Arteria innominata.
　3. Art. subclavia dextra.
　4. Art. carotis communis dextra.
　5. Art. carotis interna.
　6. Art. carotis externa.
　7. Art. thyreoidea superior.
　8. Art. maxillaris externa v. facialis.
　9. Art. occipitalis.
　10. Art. auricularis posterior.
　11. Art. temporalis.
　12. Artt. and vv. pulmonales.
　13. Art. and ven. intercostalis.
　14. Aorta descendens abdominalis (with plexus aorticus inferior).
　15. Art. cœliaca with plexus cœliacus.
　16. Art. renalis with plexus renalis.
　17. Art. mesenterica superior with plexus mesentericus superior.
　18. Art. mesenter. inferior with plexus mes. inf.
　19. Art. iliaca communis.
　20. Plexus hypogastricus superior.
　21. Plexus hæmorrhoidalis.
　22. Plexus vesicalis.
　23. Plexus prostaticus.
　24. Plexus hypogastricus inferior.
　25. Artt. phrenicæ inferiores with plexus phrenicus.
　26. Plexus gastricus magnus.
　27. Art. splenica with plexus splenicus.
　28. Art. hepatica with plexus hepaticus.
　29. Plexus solaris, with ganglia semilunaria.
　30. Ganglion lumbale.
　31. Ganglion sacrale.
　32. Ganglion thoracicum (1).
　33. Ganglion thoracicum (7).
　34. Nerv. splanchnicus major.
　35. Nerv. splanchnicus minor.
　36. Plexus thoracicus superior.
　37. Ganglion cervicale inferior.
　38. Ganglion cervicale medium.
　39. Ganglion cervicale superior.
　40. Plexus nervorum mollium.
　41. Nerv. auricularis anterior.
　42. Nerv. auricularis posterior.
　43. Nerv. facialis and pes anserinus.
　44. Nerv. occipitalis minor and nerv. auricularis superior.
　45. Nerv. accessorius Willisii.
　46. Plexus cervicalis.
　47. Nerv. vagus.
　48. Nerv. recurrens.
　49. Nerv. phrenicus.
　50. Plexus brachialis.
　51. Plexus lumbalis.
　52. Plexus sacralis.
　53. Nerv. intercostalis.
　54. Plexus œsophageus.
　55. Plexus pulmonalis.
　56. Plexus pharyngeus.

Tab XXV

Fig 7.

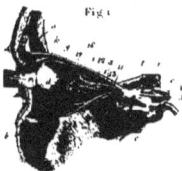

Fig 1.

Fig 2.

Fig 6.

Fig 3.

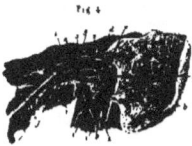

Fig 4.

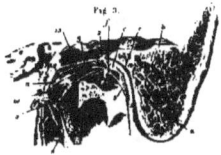

Fig 5.

PLATE XXXI.

THE EAR, EXTERNAL AND INTERNAL.
THE EYE, ORBIT, AND LACHRYMAL APPARATUS.

Fig. 1.

External Ear and Auricular Muscles—Left.

a Helix.
b Spina v. processus acutus helicis.
c Anthelix.
d Crus superius anthelicis.
e Crus inferius anthelicis.
f Fossa innominata v. triangularis.
g Scapha v. fossa navicularis.
h Tragus.
i Antitragus.
k Incisura auriculæ.
l Concha auris.
  Opening of meatus auditorius externus.
m Mscl. attolens auriculæ.
n M. attrahens auriculæ.
o Mm. retrahentes auriculæ
p M. helicis major.
q M. helicis minor.
r M. tragicus.
s M. antitragicus.

Fig. 2.

Posterior view of Ear (Left), with Auricular Muscles.

a Helix.
b Fossa navicularis v. scapha.
c Anthelix.
d Concha.
e Meatus auditorius extern. cartilagineus.
f Mscl. attollens auriculæ.
g M. attrahens auriculæ.
h Mm. retrahentes auriculæ.
i M. transversus auriculæ.
k M. obliquus auriculæ.

Fig. 3.

Vertical Section of Temporal Bone, showing Internal Ear, Tympanum, Cochlea, and Semicircular Canals—Left side.

a Pars squamosa ossis temporum.
b Pars petrosa ossis temporum.
c Cavitas tympani.
d Membrana tympani.
e Caput mallei.
f Incus.
g Processus brevis incudis.
h Processus longus incudis.
i Capitulum stapedis.
k Crus anterius stapedis.
l Crus posterius stapedis.
m Basis (in the fenestra ovalis) stapedis.
n Mscl. tensor tympani.
o M. stapedius.
p Meatus auditorius internus.
q Cochlea.
r Lamina spiralis.
s Modiolus.
t Scala tympani.
u Scala vestibuli.
v Vestibulum.
w Canalis semicircularis posterior.
x Tuba Eustachii.
y Cellulæ mastoideæ.
z Canalis Fallopii (pro nerv. faciali).

Fig. 4.

The cavity of the Tympanum, showing the Ossicular or Small Bones of Hearing—Left side.

a Inner wall of tympanum.
b Fenestra rotunda.
c Promontorium.
d Caput mallei.
e Processus longus mallei.
f Manubrium mallei.
g Processus brevis v. obtusia mallei.
h Corpus incudis.
i Processus brevis incudis.
k Processus longus incudis.
l Stapes.
m Eminentia papillaris.
n Tendo mscl. stapedii.
o Lig. capituli mallei v. superius.
p Lig. incudis superius.

q Lig. processus brevis incudis.
d Entrance to mastoideon cells.
s Tuba Eustachii pars ossea.
t Tuba Eustachii pars cartilaginea.
u Semicanalis pro m. tensore tympani.

Fig. 5.

The Internal organs of Hearing connected conjointly, without Bony Structures.

a Auricula externa (anterior margin).
b Meatus auditor. externus.
c Membrana tympani.
d Malleus.
e Processus longus mallei.
f Manubrium mallei.
g Incus.
h Processus brevis incudis.
i Processus longus incudis.
k Osiculum orbiculare Sylvii.
l Stapes.
m Vestibulum.
n Canalis semicircularis superior.
o Posterior canalis semicircularis.
p Inferior canalis semicircularis.
q Cochlea.
r Cupola cochleæ.

Fig. 6.

The Ossicula, or Bones of Tympanum, articulated with each other—Left Ear.

A. malleus.
a Capitulum.
b Superficies articularis capituli.
c Collum.
d Processus longus.
e Processus brevis.
f Manubrium mallei.

B. incus.
a Corpus incudis.
b Cavitas glenoidalis pro capitulo mallei.
c Processus brevis.
d Processus longus.
e Osiculum orbiculare v. lenticulare Sylvii.

C. stapes.
a Capitulum stapedis.
b Crus anterius.
c Crus posterius.
d Basis.
e Sulcus stapedis.

Fig. 7.

Incus—Left Ear.

a Corpus incudis.
b Cavitas glenoidalis pro capitulo mallei.
c Processus brevis.
d Processus longus.
e Osiculum orbiculare v. lenticulare Sylvii.

Fig. 8.

Malleus—Left Ear.

a Capitulum.
b Superficies articularis capituli.
c Collum.
d Processus longus.
e Processus brevis.
f Manubrium mallei.

Fig. 9.

Stapes.

a Capitulum stapedis.
b Crus anterius.
c Crus posterius.
d Basis.
e Sulcus stapedis.

Fig. 10.

The Bony Labyrinth—Left Ear (external surface).

a Fenestra ovalis.
b Fenestra rotunda.

c Promontorium.
d Canalis semicircularis superior.
e Crus anterius of superior semicircular canal.
f Ampulla os- of posterior semicircular canal.
  Crus poste- lar canal.
rius.
h Canalis semicircularis posterior v. inferior.
i Crus superius of posterior semicircular canal.
k Crus inferius rior semicircular canal.
  Ampulla os- circular canal.
l sa inferior canal.
m Canalis semicircular. externus v. horizontalis.
n Crus anterius of horizontal semicircular canal.
o Ampulla ossea anterior.
p Inferior curve of cochlea.
q Third (half) curve of cochlea.
r Cupola v. apex cochleæ.

Fig. 11.

Interior of Bony Labyrinth —Left Ear.

a Vestibulum.
b Canalis semicircularis superior.
c Canalis semicircularis posterior.
d Canalis semicircularis externus.
e Cochlea.
f Lamina spiralis.
g Scala tympani v. inferior.

Fig. 12.

Interior of Labyrinth, with distribution of the Acoustic Nerve (Portio Mollis of seventh)—Left Ear.

a Cochlea.
b Cupola.
c Border of spiral plate.
d Nervus acousticus v. auditorius.
e Nerv. cochleæ.
f Nerv. vestibuli.
g Nerv. macularis major.
h Nerv. macularis minor.
i Nerv. ampullaris inferior.
k Canalis semicircularis posterior.
l Ampulla inferior.
m Canalis semicircularis superior.
n Ampulla superior.
o Common termination of superior and posterior semicircular canals.
p Canalis semicircularis externus.
q Sacculus oblongus.

Fig. 13.

Internal linings of Laby-rinth—Left Ear.

a Sacculus rotundus.
b Sacculus oblongus, alœus v. sinus comm.
c Canalis semicircularis superior.
d Ampulla membranacea superior.
e Canalis semicircularis externus.
f Ampulla membranacea externa.
g Canalis semicircularis posterior.
h Common termination of superior and posterior semicircular canals.
i Ampulla membranacea inferior.
k First winding of spiral plate, or
l Second winding lamina
m Third (half) spiralis.
  winding
n Hamulus laminæ spiralis.
o Scyphus.
p Zonula ossea laminæ spiralis.
q Zonula membranacea lamina spiralis v. zona Valsalvæ.

Fig. 14.

The Osseous or Bony Cochlea—Left Ear—showing its internal arrangement.

a Basis Cochleæ.
b Cupula v. apex cochleæ.
c First winding of
d Second winding cochleæ.
e Third (half) winding
f Outer wall of cochleæ.
g Inner wall
h Basis modioli.
i Modiolus.
k Apex modioli (columella).
l Canales of modiolus with cupola.
m Lamina spiralis ossea.
n Scala tympani v. inferior.
o Scala vestibuli v. superior.
p Termination of semicircular canals (scyphus).

Fig. 15.

The Osseous or Bony Cochlea—Left Ear—cut through its centre.

p As in Fig. 14.
q Canalis centralis modioli.

Fig. 16.

Facial surface of Eyelids and Eye—Left side—with Blood-vessels and Nerves.

a Musculi orbicularis oculi strutum externum.
b Mscl. orbicularis stratum internum.
c Palpebra superior.
d Lig. palpebralis internum.
e Mscl. frontalis.
f M. zygomaticus major.
g M. zygomaticus minor.
h M. levator labii superioris proprius.
i M. levator labii superioris alæque nasi.
k M. masseter.
1 Arteria angularis.
2 Art. ophthalmica.
3 Art. dorsalis nasi.
4 Art. temporalis.
5 Ramus frontalis art. temporalis.
6 Vena temporalis.
7 Ven. facialis anterior.
8 Ven. frontalis.
9 Nerv. facialis rami zygomatici.
10. Nerv. auricularis anterior.
11. Nerv. infraorbitalis.
12. Nerv. supraorbitalis.
13. Nerv. supratrochlearis.
14. Nerv. infratrochlearis.
15. Nerv. ethmoidalis.

Fig. 17.

Anterior and Internal Muscles of Eye—right.

a Mscl. orbicularis stratum externum.
b M. obliquus oculi inferior.
c M. obliquus oculi superior.
d Trochlea m. obliq. super.
e M. occipito-frontalis.
f M. levator labii superioris alæque nasi.
g M. levator labii superioris.
h M. zygomaticus minor.
i Adipose tissue surrounding globe of eye.
k Bulbus oculi (eyeball).
l M. rectus oculi superior.
m M. rectus oculi externus.
n M. rectus oculi internus.
o M. rectus oculi inferior.
p M. rectus oculi externus.

Fig. 18.

Lachrymal Apparatus of Eye and Eyelids—Left side.

a Margo supraorbitalis (ossis frontalis).
b Margo infraorbitalis.
c Bulbus oculi (eyeball).
d Palpebra superior.

e Mscl. levator palpebræ superioris.
f Glandula lacrymalis.
g Openings of Meibomian glands.
h Glandulæ Meibomianæ.
i Puncta lacrymalia.
k Canaliculi lacrymalea.
l Saccus lacrymalis.
m Canalis lacrymalis, or nasal duct.

Fig. 19.

Upper surface of Orbital Cavities (the osseous structures being removed), with Muscles and Nerves.

a Cerebral surface of orbital plate of os frontis.
b Orbita galli.
c Lamina cribrosa.
d Corpus ossis sphenoidei.
e Sella turcica.
f Ala magna ossis sphenoidei.
g Outer wall of orbit.
h Bulbus oculi.
i Glandula lacrymalis.
k Mscl. rectus oculi superior.
l M. rectus oculi internus.
m M. obliquus oculi superior.
n Trochlea m. obliqui super.
o M. levator palpebræ superioris.
p M. rectus oculi externus.
q M. occipito-frontalis.
1. Arteria ophthalmica.
2. Tractus opticus.
3. Commissura nervorum opticorum.
4. Nerv. opticus.
5. Nerv. oculomotorius (third nerve).
6. Nerv. trigeminus (fifth nerve).
7. Ganglion semilunare v. Gasseri.
8. Ramus oph-
   thalmicus
9. Ram. maxilla- trige-
   ris superior minal. Nervi
10. Ram. maxilla-   trige-
    ris inferior.   mini.
11. Nerv. abducens (sixth nerve).
12. Nerv. trochlearis (fourth nerve).
13. Ramus frontalis nervi ophthalmici.
14. Nerv. supraorbitalis.
15. Nerv. lacrymalis.

Fig. 20.

Vertical section of Orbit and Globe of Eye, through centre, from before backwards.

a Os frontis.
b Os maxillare superius.
c Adipose tissue.
d Mscl. frontalis.
e Palpebra superior.
f Palpebra inferior.
g Mscl. obliquus oculi inferior.
h M. rectus oculi inferior
i M. rectus oculi externus.
k M. rectus oculi superior.
l M. levator palpebræ superioris.
m Nervus opticus.
n Conjunctiva palpebræ.
o Reflection of conjunctiva from inner surface of eyelids to globe of eye.
p Conjunctiva sclerotica (bulbi).
q Conjunctiva cornea.
r Cornea.
s Membrane of aqueous humour lining anterior chamber.
t Camera oculi an-
   terior   with
u Camera oculi humor
   posterior   aqueus.
v Sinus venosus iridis.
w Tunica sclerotica.
x Lens crystallina.
y Corpus ciliare.
z Corpus vitreum.
a Tunica retina.
β Tunica Choroidea.

Tab. XXXI

Fig. 2

Fig. 3

Fig. 4

Fig. 1

Fig. 14

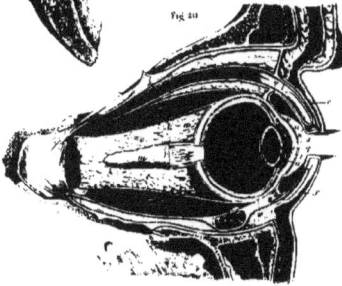

Fig. 20

Fig. 11

Fig. 15

Fig. 10

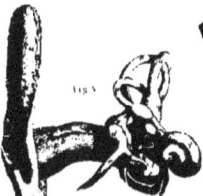

Fig. 5

Fig. 19

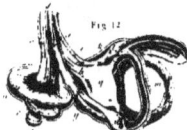

Fig. 12

Fig. 6

Fig. 9

Fig. 7

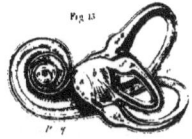

Fig. 13

Fig. 18

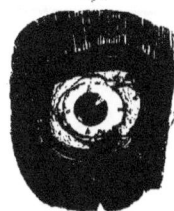

Fig. 17

Fig. 16

PLATE XXXII.

THE LARYNX, WITH SECTIONS OF NASAL CAVITY AND MOUTH, THE TONGUE, AND FAUCES.

Fig. 1.
Anterior view of Larynx and Os Hyoidea.

a   Corpus ossis hyoidei.
b   Cornu majus ossis hyoidei.
c   Cornu minus ossis hyoidei.
d   Lig. thyro-hyoideum medium.
e   Lig. thyro-hyoideum laterale.
f   Cartilago thyroidea.
g   Pomum Adami.
h   Cornua superiora v. majora.
i   Cartilago cricoidea.
k   Lig. crico-thyroideum medium.
l   Lig. crico-thyroideum laterale.
m.n Annulus cartilagineus tracheæ (1) et (2).

Fig. 2.
Posterior view of Larynx.

a   Corpus ossis hyoidei.
b   Cornu minus ossis hyoidei.
c   Cornu majus ossis hyoidei.
d   Lig. thyro-hyoideum laterale.
e   Cornu superius of thyroid cartilage.
f   Cartilago thyroidea.
g   Cartilago cricoidea.
h   Cartilago arytænoidea.
i   Cartilago Santoriniana.
k. l Epiglottis.

Fig. 3.
Arteries of Tongue, Palate, and Nasal Cavity.

a   Sinus frontalis.
b   Sinus sphenoidalis.
c   Superior spongy bone.
d   Middle spongy bone.
e   Inferior spongy bone.
f   Palatum durum.
g   Uvula.
h   Dorsum linguæ.
i   Musculus styloglossus.
k   M. genioglossus.
l   M. geniohyoideus.
m   M. hyoglossus.
n   M. mylohyoideus.
o   M. sternohyoideus.
p   M. geniohyoideus sinister.
q   M. omohyoideus.
r   M. stylopharyngeus.
s   Cartilago thyroidea.
t   Maxilla inferior.

  1. Arteria carotis communis.
  2. Art. carotis interna.
  3. Art. carotis externa.
  4. Art. thyroidea superior.
  5. Art. laryngea superior.
  6. Art. lingualis.
  7. Ramus hyoideus of art. lingualis.
  8. Art. dorsalis linguæ.
  9. Art. maxillaris externa v. facialis.
  10. Art. sublingualis.
  11. Art. ramina v. profunda linguæ.
  12. Rami art. sphenopalatinæ.
  13. Art. palatina superior.
  14. Art. ethmoidalis anterior.
  15. Art. ethmoidalis posterior.

Fig. 4.
Upper or dorsal surface of Tongue.

a   Radix v. basis linguæ.
b   Apex linguæ.
c   Margines linguæ.
d   Dorsum linguæ.
e   Glandulæ mucosæ.
f   Foramen cæcum.
g   Papillæ vallatæ v. truncatæ.
h   Papillæ fungiformes v. lenticulares.
i   Papillæ minores (conicæ et filiformes).
k   Fimbriæ v. rugæ transversæ.
l   Plica glosso-palatina.
m   Epiglottis.
n   Lig. glosso-epiglotticum v. frænulum epiglottidis.

Fig. 5.
Inferior or concave surface of Tongue divided by longitudinal section.

a   Os hyoideum.
b   Apex linguæ.
c   Sublingual follicles.
d   Mscl. geniohyoideus.
e   M. thyrohyoideus.
f   M. styloglossus.
g   M. hyoglossus.
h   M. lingualis.
i   M. genioglossus.
k   Glandula submaxillaris.
l   Nervus hypoglossus.
m   Nerv. lingualis.
n   Ganglion maxillare.
o   Radix mollis v. sympathica.
p   Radix longa v. motoria.
q   Nervuli ad ductum Whartonianum.
r   Connecting filament between nervi hypoglossi.

Fig. 6.
Nasal Bones and Cartilages—anterior view.

a   Processus nasalis ossis frontis.
b   Processus nasalis ossis maxillaris superioris.
c   Ossa nasi.
d   Processus alveolaris maxillæ superioris.
e   Spina nasalis anterior.
f   Cartilagines nasi laterales superiores.
g   Cartilagines nasi laterales inferiores.
h   Cartilago alæ nasi propria.
i   Cartilago lateralis inferior posterior.
k   Cartilagines alarum nasi minores.
l   Cartilago septi narium.

Fig. 7.
Vertical and transverse section of Nasal Cavity and Antrum Highmorianum through centre.

a   Orbita.
b   Lamina cribrosa ossis ethmoidei.
o   Septum nasi.
d   Superior spongy bone.
e   Middle spongy bone.
f   Inferior spongy bone.
g   Cellulæ ethmoidales.
h   Sinus maxillaris v. antrum Highmori.
i   Palatum durum.

Fig. 8.
Vertical section of Nasal Cavity, Mouth and Larynx, through centre.

a   Sinus frontalis ossis frontis.
b   Os nasi.
c   Lamina cribrosa ossis ethmoidei.
d   Sinus sphenoidalis.
e   Palatum durum.
f   Canalis incisivus.
g   Maxilla inferior.
h   Os hyoideum.
i   Superior spongy bone.
k   Middle spongy bone.
l   Inferior spongy bone.
m   Ostium pharyngeum tubæ Eustachii.
n   Palatum molle (uvula).
o   Pharynx.
p   Lingua (tongue).
q   Mscl. genioglossus.
r   M. geniohyoideus.
s   Epiglottis.
t   Lig. thyro-arytænoideum superius.
u   Lig. thyro-arytænoideum inferius.
v   Ventriculus Morgagnii.
w   Cartilago thyroidea.
x   Cartilago arytænoidea.
y   Cartilago cricoidea.
z   Œsophagus.

Fig. 9.
Vertical section of Nasal Cavity, Mouth and Larynx, to left side of centre.

a   Sinus frontalis ossis frontis.
b   Os nasi.
c   Sinus sphenoidalis.
d   Lamina cribrosa ossis ethmoidei.
e   Os maxillare superius.
f   Canalis incisivus.
g   Palatum durum.
h   Palatum molle v. velum palatinum.
i   Lingua.
k   Os maxillare inferius.
l   Mscl. genioglossus.
m   Septum nasi.
n   Posterior nares.
o   Pharynx.
p   Tonsilla.
q   Arcus pharyngo-palatinus.
r   Epiglottis.
s   Frænulum epiglottidis.
t   Posterior wall of larynx.
u   Œsophagus.
v   Trachea.
w   Nervus olfactorius.
x   Nerv. ethmoidalis v. nasalis anterior.
y   Nerv. nasopalatinus Scarpæ.
z   Ganglion incisivum.

Tab. XXXII.

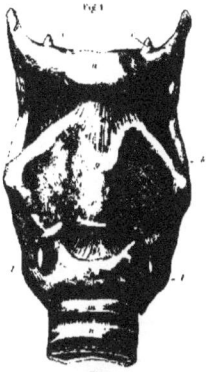

Fig. 1.

Fig. 2.

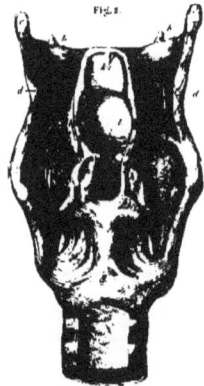

Fig. 3.

Fig. 4.

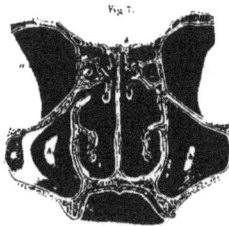

Fig. 7.

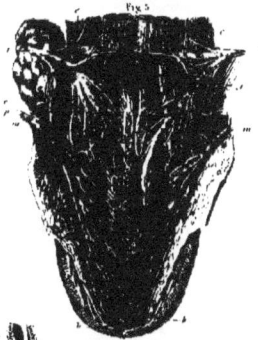

Fig. 5.

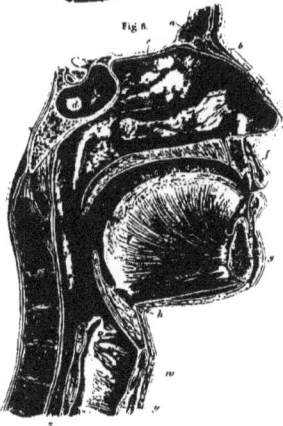

Fig. 8.

Fig. 6.

Fig. 9.

PLATE XXXIII.

VISCERA OF THORAX, ABDOMEN, AND PELVIS

(Anterior View).

Fig. 1.

The Thoracic Parietes with Viscera enclosed, the Abdomen and Abdominal Viscera in natural position.

a Clavicula.
b Sternum.
c First rib.
d Tenth rib.
e Cartilagines costarum.
f Os ilium.
g Os pubis.
h Musculus pectoralis minor.
i Musculi intercostales interni.
k Macl. triangularis sterni.
l M. subscapularis.
m M. latissimus dorsi.
n Mm. abdominales (m. obliquus externus and internus, transversalis).
o M. sartorius.
p M. rectus femoris.
q M. tensor fasciæ latæ.
r M. adductor femoris longus.
s M. pectineus.
t Ligamentum Poupartii.
u Lig. suspensorium penis.
v Penis.
w Spermatic cord.
x Divided margin of m. obliquus externus.
y Fascia transversalis.
z Inferior pillar of external abdominal ring (annulus abdominalis).
  1. Arteria axillaris.
  2. Vena axillaris.
  3. Art. and ven. mammaria interna.
  4. Artt. intercostales anteriores superiores.
  5. Artt. intercostales anteriores inferiores.
  6. Rami sternales of art. mammaria interna.
  7. Plexus brachialis.
  8. Art. and ven. transversa scapulæ with nerv. suprascapularis.
  9. Artt. intercostales posteriores.
  10. Nervi intercostales.
  11. Art. cruralis.
  12. Vena cruralis.
  13. Art. and ven. epigastrica.
  14. Vena saphæna magna.
  15. Art. and ven. circumflexa ilium.
  16. Nervus cruralis.
  17. Ramus anterior nervi obturatorii.
  18. Nerv. cutaneus femoris anterior externus.
  19. Ramus cutaneus nervi iliohypogastrici.
  20. Nerv. lumbo-inguinalis.
  I. Pleura costalis.
  II. Pulmo sinister.
  III. Mediastinum anticum.
  IV. Pleura phrenica.
  V. Diaphragm.
  VI. Peritonæum.
  VII. Fossa inguinalis externa.

VIII. Peritoneal coat of
IX. Vesica urinaria.
X. Ligamentum suspensorium hepatis.
XI. Umbilicus.
XII. Ligamentum teres hepatis (obliterated vena umbilicalis).
XIII. Ligg. vesicæ lateralia (obliterated artt. umbilicales).
XIV. Lig. vesicæ medium (obliterated urachus).
XV. Stomach.
XVI. Lobus dexter hepatis, with gall bladder.
XVII. Lobus sinister hepatis, with gall bladder.
XVIII. Colon transversum.
XIX. Intestinum cæcum.
XX. Intestinum jejunum and ileum.
XXI. Colon descendens.
XXII. S. romanum v. flexura iliaca.
XXIII. Intestinum rectum.

Fig. 2.

The Lungs *in situ*, the Deeper Abdominal Viscera, Small Intestines being removed.

a Clavicula.
b First rib.
c Eleventh rib.
d Crista Ilii.
e Musculus psoas major.
f Macl. iliacus internus.
g M. rectus femoris.
h M. glutæus medius.
i M. vastus externus.
k M. obturator externus.
l Lig. obturatorium.
m M. adductor magnus.
n M. adductor brevis.
o M. adductor longus.
p M. gracilis.
q M. pectineus.
r M. tensor fasciæ latæ.
s M. sartorius.
t M. cruralis.
u Neck of femur.
v Trochanter major.
  1. Arteria cruralis.
  2. Vena cruralis.
  3. Art. and ven. epigastrica superficialis.
  4. Art. and ven. profunda femoris.
  5. Art. and ven. circumflexa femoris externa.
  6. Nervus obturatorius.
  I. Lobus superior of right lung.
  II. Lobus medius of right lung.
  III. Lobus inferior of right lung.
  IV. Lobus superior of left lung.
  V. Lobus inferior of left lung.
  VI. Pleura.
  VII. Mediastina antica.
  VIII. Diaphragm.

IX. Œsophagus.
X. Stomach.
XI. Spleen.
XII. Left lobe of liver, a portion of left extremity being removed.
XIII. Right lobe of liver.
XIV. Gall bladder.
XV. Lig. suspensorium hepatis.
XVI. Duodenum.
XVII. Jejunum.
XVIII. Mesenterium.
XIX. Cæcum.
XX. Processus vermiformis.
XXI. Colon transversum.
XXII. Flexura coli dextra.
XXIII. Colon transversum.
XXIV. Flexura coli sinistra.
XXV. Colon descendens.
XXVI. S. romanum v. flexura iliaca (sigmoid flexure of colon).
XXVII. Rectum.
XXVIII. Peritonæum.
XXIX. Ileum (divided).
XXX. Penis (corpora cavernosa penis and spongiosum urethræ).

Fig. 3.

Under or Concave Surface of Liver.

I. Lobus dexter.
II. Lobus sinister.
III. Lobulus quadratus.
IV. Lobulus Spigelii.
V. Porta hepatis.
VI. Gall bladder.
a Anterior margin.
b Posterior margin.
c Lig. suspensorium hepatis.
d Lig. teres hepatis (in fossa umbilicalis).
e Vena cava inferior.
f Fossa ductus venosi.
g Vena portæ.
h Arteria hepatica.
i Ductus choledochus.
k Ductus cysticus.
l Ductus hepaticus.

Fig. 4.

Internal arrangement of Hepatic Blood-vessels, the Liver being divided transversely.

I.—V. as in Fig. 3.
a—1 as in Fig. 3.
m Ductus venosus.
n Art. cystica.
o Fundus of gall bladder.
p Collum of gall bladder.
q Venæ hepaticæ.

Tab.XXXIII

Fig. 1.

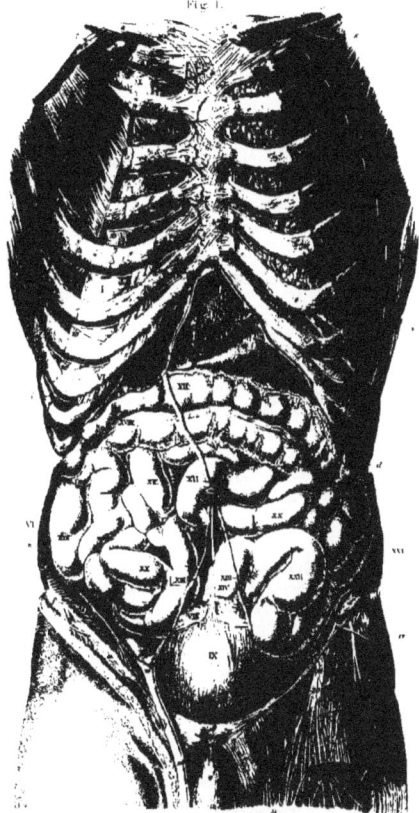

Fig. 2.

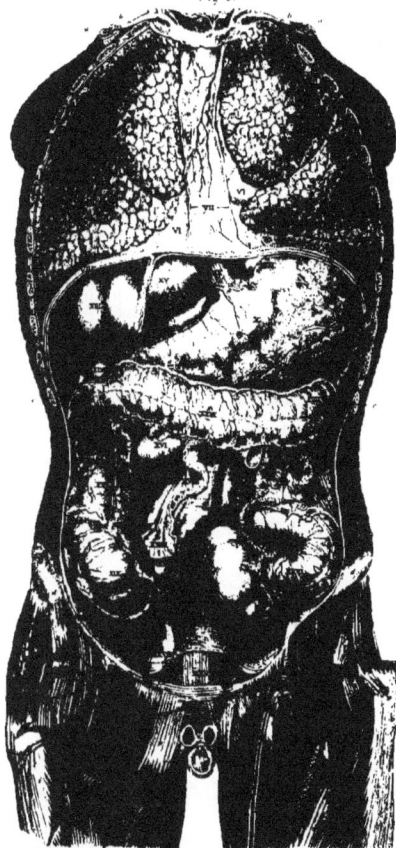

Fig. 3.

Fig. 4.

J. Renner in Leipzig

PLATE XXXIV.

PRINCIPAL ORGANS OF DIGESTION, WITH DEEPER BLOOD-VESSELS OF ABDOMINAL VISCERA.

Fig. 1.

Liver, Stomach (raised upwards), Pancreas, and Spleen *in situ*, with their principal Arteries.

a   Crura interna diaphragmatis.
b   Stomach.
c   Curvatura major.
d   Curvatura minor.
e   Saccus cæcus v. fundus.
f   Cardiac extremity.
g   Pylorus.
h   Duodenum.
i   Pancreas.
k   Caput pancreatis.
l   Cauda pancreatis.
m   Spleen.
n   Lobus sinister hepatis.
o   Lobulus quadratus hepatis.
p   Lobus dexter hepatis.
q   Gall bladder.
r   Ductus choledochus.
s   Vena portæ.
   1. Aorta descendens abdominalis.
   2. Art. phrenica inferior.
   3. Art. cœliaca.
   4. Art. coronaria ventriculi sinistra.
   5. Art. splenica.
   6. Art. hepatica.
   7. Art. coronaria ventriculi dextra.
   8. Art. gastro-duodenalis.
   9. Art. gastro-epiploica dextra.
  10. Art. pancreatico-duodenalis.
  11. Artt. v. rami breves ventriculi.
  12. Art. gastro-epiploica sinistra.
  13. Art. cystica.
  14. Art. mesenterica superior.

Fig. 2.

Under surface of Liver (raised upwards), with Stomach *in situ*, and principal Blood-vessels.

a   Crura interna diaphragmatis.
b   Stomach (anterior surface).
c   Curvatura major.
d   Curvatura minor.
e   Saccus cæcus v. fundus.
f   Cardiac extremity.
g   Pylorus.
h   Duodenum.
i   Spleen.
k   Lobus sinister hepatis.
l   Lobulus quadratus.
m   Lobulus Spigelii.
n   Lobus dexter hepatis.
o   Gall bladder.
p   Ductus cysticus.
q   Ductus hepaticus.
r   Ductus choledochus.

Fig. 3.

Small Intestines (Jejunum and Ileum), Mesentery, and Mesenteric Vessels.

a   Omentum (raised and thrown back).
b   Intestinum cæcum.
c   Colon ascendens v. dextrum.
d   Colon transversum.
e   Commencement of jejunum.
f   Intestinum jejunum.
g   Intestinum ileum.
h   Mesenterium.
i   Mesocolon dextrum.
   1. Art. mesenterica v. mesraica superior.
   2. Vena mesenterica major.
   3. Artt. et vv. jejunales; } artt. et vv. in-
   4. Artt. et vv. ileæ;   } testinales.
   5. Art. et ven. ileo-colica.
   6. Art. et ven. colica dextra.

Fig. 4.

Large Intestines, with principal Blood-vessels.

a   Divided end of jejunum.
b   Divided end of ileum.
c   Mesenterium (divided) with principal blood-vessels.
d   Intestinum cæcum.
e   Colon ascendens v. dextrum.
f   Colon transversum.
g   Colon descendens v. sinistrum.
h   Flexura iliaca (sigmoid flexure).
i   Commencement of rectum.
k   Mesocolon transversum.
l   Mesocolon dextrum.
m   Mesocolon sinistrum.
n   Mesocæcum.
   1. Art. mesenterica superior.
   2. Vena mesenterica major.
   3. Art. et ven. colica media.
   4. Art. et ven. colica dextra.
   5. Art. et ven. ileo-colica.
   6. Art. mesenterica inferior.
   7. Vena mesenterica minor.
   8. Art. et ven. colica sinistra.
   9. Art. et ven. hæmorrhoidalis interna.

Fig. 5.

View of posterior surface of the deep Viscera of Abdomen and Pelvis, with principal Blood-vessels.

a   Tenth dorsal vertebra.
b   Last rib.
c   Os ilium.
d   Diaphragm.
e   Glandula suprarenalis.
f   Right kidney.
g   Left kidney.
h   Sigmoid flexure of colon.
i   Colon ascendens and cæcum.
k   Rectum.
   1. Aorta descendens abdominalis.
   2. Vena cava inferior.
   3. Art. and ven. renalis.
   4. Art. iliaca (communis).
   5. Ven. iliaca (communis).
   6. Art. iliaca interna.
   7. Ven. iliaca interna.
   8. Ven. iliaca externa.

Fig. 6.

View of posterior surface of the superficial Viscera of Abdomen and Blood-vessels.

a   Vena cava inferior.
b   Liver.
c   Spleen.
d   Pancreas.
e   Caput pancreatis.
f   Cauda pancreatis.
g   Duodenum.
h   Ileum.
i   Cæcum.
k   Colon ascendens.
l   Colon descendens.
m   Sigmoid flexure of colon descendens.
n   Rectum.
   1. Art. cœliaca.
   2. Art. splenica.
   3. Art. hepatica.
   4. Art. mesenterica superior.
   5. Art. mesenterica inferior.
   6. Art. and ven. hæmorrhoidalis interna.
   7. Art. colica sinistra.
   8. Ven. colica sinistra.
   9. Ven. mesenterica minor.
  10. Ven. splenica.
  11. Ven. mesenterica magna.
  12. Art. and ven. ilio-colica.
  13. Art. and ven. colica dextra.

Tab. XXXIV.

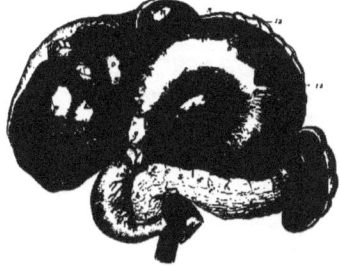

Fig. 2.

Fig. 1.

Fig. 4.

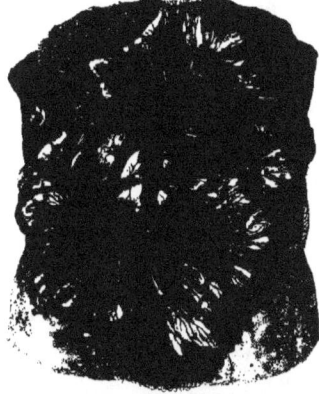

Fig. 3.

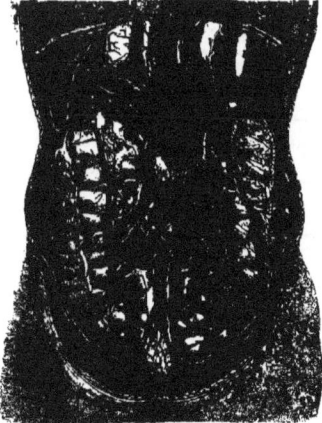

Fig. 6.

Fig. 5.

PLATE XXXV.

THORACIC AND ABDOMINAL VISCERA, WITH PRINCIPAL VESSELS, NERVES, AND LYMPHATICS.

Fig. 1.

Anterior View.

a Clavicula.
b First rib.
c Glandula thyroidea.
d Trachea.
e Bronchus dexter.
f Bronchus sinister.
g Dorsal spine.
h Pulmo dexter.
i Mediastinum posticum.
k Diaphragm.
l Stomach.
m Spleen.
n Lobus sinister hepatis.
o Lobus dexter hepatis.
p Colon ascendens.
q Mesenterium.
r Jejunum and ileam.
s Gall bladder.
t Ligamentum suspensorium hepatis.
1. Arcus aortæ.
2. Aorta descendens thoracica.
3. Arteria subclavia.
4. Art. carotis communis.
5. Art. innominata.
6. Artt. and vv. intercostales.
7. Vena cava superior.
8. Ven. innominata v. jugularis communis dextra.
9. Ven. innominata v. sinistra.
10. Ven. subclavia.
11. Ven. jugularis interna.
12. Ven. azygos.
13. Ven. demiazygos.
14. Art. mesenterica.
15. Ven.-magna.
16. Artt. and vv. jejunales et ileam.
17. Ductus thoracicus
18. Ductus dexter v. minor.
19. Glandulæ bronchiales.
20. Glandulæ pulmonicæ.
21. Glandulæ jugulares profundæ.
22. Glandulæ axillares.
23. Glandulæ intercostales.
24. Plexus mesentericus with glandulæ mesentericæ.

Fig. 2.

Posterior view.

a Corpus of first dorsal vertebra.
b Processus spinosus of first dorsal vertebra.
c First rib.
d Scapula.
e Medulla spinalis.
f Œsophagus.
g Trachea.
h Apex of right lung.
i Pleura costalis.

k Diaphragm.
l Heart.
m Bronchus sinister.
n Kidney.
o Pelvis renalis.
p Ureter.
q Glandula suprarenalis.
r Peritonæum.
s Intestinum rectum.
t Musc. sphincter ani externus.
u Musc. levator ani.
v Great sacro-sciatic ligament.
w Musc. pyriformis.
x Os ilii.
y Musc. psoas major.
z Mm. glutæi.
1. Arcus aortæ.
2. Aorta descendens thoracica.
3. Aorta descendens abdominalis.
4. Art. iliaca communis.
5. Art. and ven. iliaca interna.
6. Art. and ven. iliaca externa.
7. Art. and ven. sacra media.
8. Art. innominata.
9. Art. subclavia.
10. Art. carotis communis.
11. Art. and ven. mammaria interna.
12. Art. ven. and nerv. intercostales.
13. Art. and ven. renalis (with rami. suprarenalis).
14. Art. and ven. spermatica interna.
15. Art. and ven. hæmorrhoidalis interna.
16. Art. and vv. hæmorrhoidales mediæ.
17. Art. and ven. pudenda communis.
18. Art. and ven. ischiadica.
19. Art. and ven. glutæa superior.
20. Vena subclavia.
21. Ven. cava superior.
22. Ven. azygos.
23. Ven. demiazygos.
24. Ven. lumbalis (1 and 2).
25. Ven. cava inferior.
26. Ven. iliaca communis.
27. Ductus thoracicus.
28. Receptaculum chyli.
29. Glandulæ lumbales.
30. Glandulæ intercostales.
31. Glandulæ mediastini postici.
32. Nerv. intercostalis (1).
33. Ganglion thoracicum (1).
34. Nerv. vagus.
35. Nerv. recurrens vagi.
36. Nerv. phrenicus.
37. Pars thoracica (with ganglia thoracica) nerv. sympathici.
38. Nerv. splanchnicus major and minor.
39. Nerv. intercostalis (12).
40. Nerv. lumbalis (1).
41. Nerv. cutaneus femoris anterior externus.
42. Nerv. cruralis.
43. Nerv. obturatorius.
44. Ganglion lumbale nerv. sympathici.
45. Plexus ischiadicus.
46. Nervi sacrales.

Fig. 3.

Principal Chylopoietic Viscera, Blood-vessels, and Ducts.

a Lobus sinister of liver, under or concave surface.
b Lobus quadratus of liver, under or concave surface.
c Lobus dexter of liver, under or concave surface.
d Lobulus Spigelii of liver, under or concave surface.
e Gall bladder.
f Ductus cysticus.
g Ductus hepaticus.
h Ductus communis choledochus.
i Pars descendens duodeni, with place of entrance of the ductus choledochus and
k Ductus pancreaticus.
l Caput pancreatis.
m Corpus pancreatis.
n Cauda pancreatis.
o Pars horizontalis inferior duodeni.
p Stomach.
q Spleen.
r Left kidney.
1. Aorta descendens abdominalis.
2. Art. axis cœliaca.
3. Art. coronaria ventriculi sinistra.
4. Art. splenica and artt. pancreaticæ.
5. Art. hepatica.
6. Art. gastro-duodenalis.
7. Art. and ven. renalis.
8. Art. and ven. mesenterica superior.
9. Vena portæ.

Fig. 4.

Posterior view of Solar Plexus and Minor Plexuses, with some of the Deep Blood-vessels.

a Diaphragm.
b Vena cava inferior (with vv. hepaticæ).
c Œsophagus.
d Stomach divided (with branches of par vagum).
e Spleen.
f Caput pancreatis.
g Cauda pancreatis.
h Kidney.
i Glandula suprarenalis.
k Ureter.
1. Aorta descendens abdominalis.
2. Art. coronaria ventriculi sinistra.
3. Art. splenica.
4. Art. hepatica (with plexus hepaticus).
5. Art. and ven. renalis (with plexus renalis).
6. Art. and ven. spermatica interna (with plexus spermaticus internus).
7. Art. mesenterica superior.
8. Vena cava inferior.
9. Plexus cœliacus v. solaris.
10. Plexus phrenicus.
11. Plexus gastricus.
12. Plexus splenicus.
13. Plexus aorticus inferior v. abdominalis.

Tab. XXXV

Fig. 1

Fig. 2

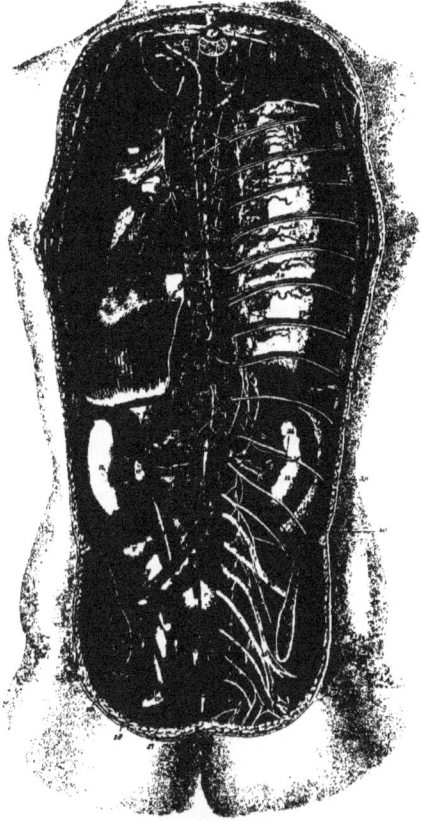

Fig. 3

Fig. 4

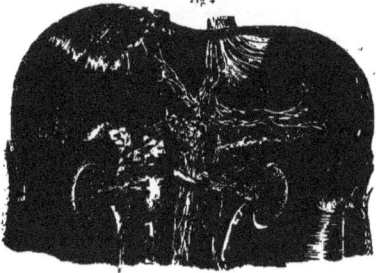

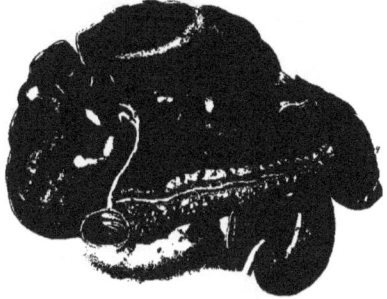

A. Krause in Leipzig

PLATE XXXVI.

GENITO-URINARY ORGANS OF MALE. AND FEMALE SUBJECTS.

Figs. 1 & 2.
Internal structure of Kidney, with Blood-vessels and Ducts.

a Cortical, cineritious or secreting surface (with tubuli contorti and Malpighian corpuscles).
b Pyramides Malpighii v. coni tubulosi.
c Papilla renalis, or mammillary process.
d Calyx renalis.
e Pelvis renalis.
f Ureter.
g Arteria renalis.
h Vena renalis.

Fig. 3.
Anterior view of cavity of Bladder laid open, with section of Genital Organs in Male subject.

a Vertex of vesica urinaria.
b Corpus of vesica urinaria.
c Fundus of vesica urinaria.
d Ureter.
e Internal opening of ureter in bladder.
f Plicæ eminentes.
g Trigone vesicale.
h Collum vesicæ.
i Prostate gland.
k Pars prostatica urethræ (with caput gallinaginis, and openings of common ejaculatory ducts).
l Pars membranacea urethræ.
m Glandula Cowperi.
n Pars spongiosa urethræ.
o Corpus spongiosum urethræ.
p Corpus cavernosum penis.
q Glans penis.
r Fossa navicularis.
s Ostium cutaneum urethræ.

Figs. 4 & 5.
Fig. 4.—Vertical section of Pelvis and Pelvic Viscera in Male, giving a lateral view of Genito-urinary Organs unopened—right side. Fig. 5.—The same parts, the interior being exposed—left side.

a Vertebra lumbalis (5).
b Os sacrum.
c Os coccygis.
d Os pubis.
e Intestinum rectum.
f Vesica urinaria.
g Ureter.
h Peritonæum.
i Penis.
k Corpus cavernosum penis (with fibres of erector penis m.).
l Urethra.
m Bulbus urethræ.
n Isthmus v. pars membranacea urethræ.
o Prostate gland and prostatic portion of urethra.

p Vesiculæ seminales.
q Vas deferens.
r Epididymis.
s Testiculus.
t Tunica vaginalis propria testiculi.
u Spermatic cord (with art. ven. and nerv. spermat. intern.).
v Glandula Cowperi.
w Musc. sphincter ani externus.
x Musc. levator ani.
y Anus.
z Lig. suspensorium penis.

Fig. 6.
Anterior view of external Male Genital Organs, the Integuments being removed on the right side.

a Dorsum penis.
b Glans penis.
c Præputium.
d Orificium cutaneum urethræ.
e Lig. suspensorium penis.
f Mons Veneris.
g External abdominal ring.
h Spermatic cord.
i Scrotum.
k Testiculus, with tunica vaginalis communis and propria.

Fig. 7.
Anterior view of Female Genito-urinary Organs and principal Blood-vessels, the Abdominal Parietes being removed.

a Os ilium.
b Crista ilii.
c Acetabulum.
d Ramus descendens ossis ischii.
e Intestinum rectum.
f Vesica urinaria.
g Ureter.
h Musc. psoas.
i Musc. iliacus internus.
k Uterus.
l Lig. uteri rotundum.
m Tuba Fallopii.
n Corpus fimbriatum and orifice of Fallopian tube.
o Lig. uteri latum.
p Ovarium.
q Lig. ovarii.
r Portio vaginalis uteri and os uteri externum.
s Vagina.
t Clitoris.
u Glans clitoridis, with præputium and frænulum.
v Labia interna, nymphæ.
w Labia externa v. majora.
x Orificium urethræ.
y Orificium vaginæ.
z Perinæum.

1. Aorta descendens abdominalis.
2. Art. iliaca communis.
3. Art. and ven. iliaca externa.
4. Art. and ven. iliaca interna.
5. Vena cava inferior.
6. Ven. iliaca communis.
7. Artt. and vv. uterinæ et vaginales.
8. Art. and ven. spermatica interna.

Fig. 8.
Vertical section of Female Genito-urinary Organs, giving lateral view of Pelvic Viscera laid open—right side.

a Last lumbar vertebra.
b Os sacrum.
c Os coccygis.
d Os pubis.
e Intestinum rectum.
f Anus.
g Vesica urinaria.
h Urachus, lig. vesicæ medium.
i Urethra.
k Orificium urethræ externum.
l Labium pudendi externum.
m Labium internum, nympha.
n Glans clitoridis.
o Corpus cavernosum clitoridis.
p Orificium vaginæ.
q Vagina.
r Portio vaginalis uteri, with os uteri externum.
s Cavum uteri.
t Canalis colli uteri.
u Fundus uteri.
v Ovarium.
w Tuba Fallopii.
x Fimbriæ v. laciniæ.
y Peritonæum.
z Peritonæum.

Fig. 9.
Transverse section of Female Genital Organs exposing the interior.

a Labia pudendi.
b Vagina (with columna rugarum posterior).
c Portio vaginalis uteri.
d Os uteri externum.
e Collum uteri.
f Canalis colli uteri with *, os uteri internum.
g Corpus uteri.
h Fundus uteri.
i Cavum uteri.
k Ostium uterinum tubæ Fallopii.
l Tuba Fallopii.
m Corpus fimbriatum, with opening of Fallopian tube.
n Ovarium.
o Ligamentum ovarii.
p Ala vespertilionis lig. uteri lati.
q Lig. uteri rotundum.
r Lig. uteri latum.

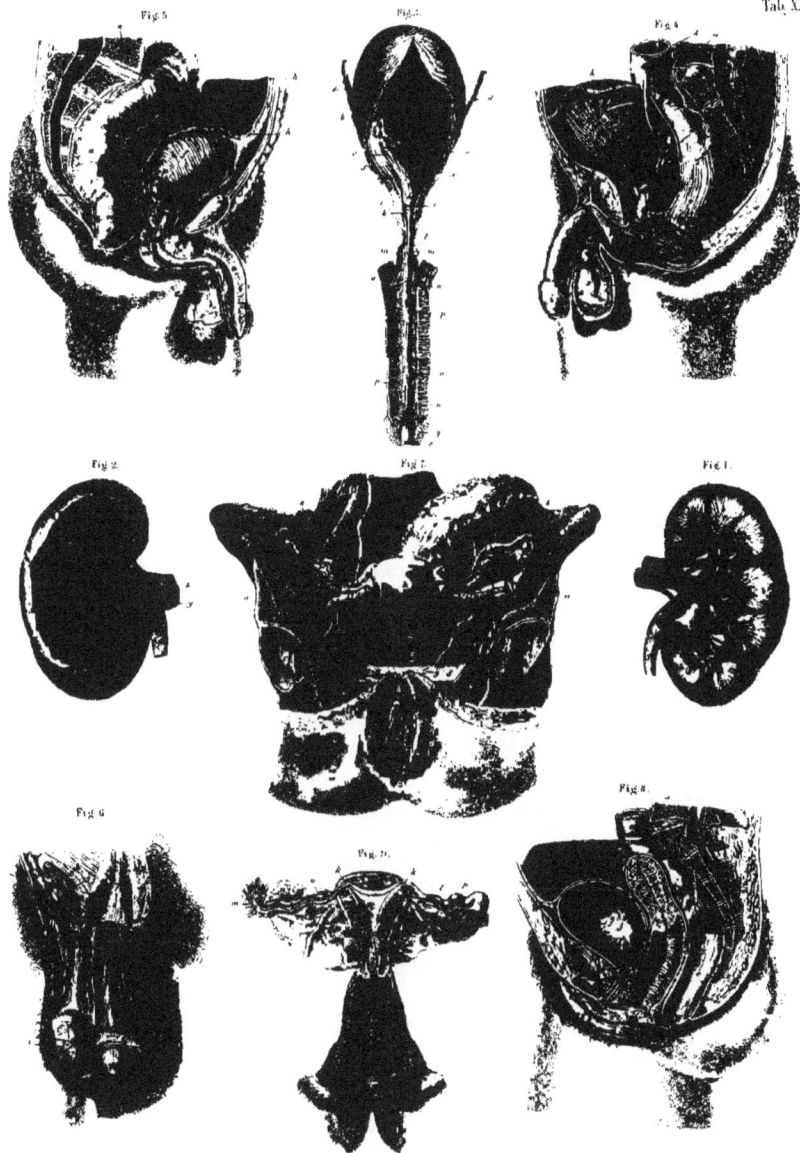

Tab. XXXVi

Fig 5. Fig 3. Fig 4.

Fig 2. Fig 7. Fig 1.

Fig 6. Fig 9. Fig 8.

A. Graefe zu Leipzig

PLATE XXXVII.

FŒTAL CIRCULATION, WITH PLACENTA AND UMBILICAL CORD, MUSCLES OF LARYNX AND LARYNGEAL INTERIOR, INTESTINUM CÆCUM LAID OPEN.—THE TEETH.

Fig. 1.

Fœtal Organization.

a Right ventricle of heart.
b L.it ventricle of heart.
c Left auricle of heart.
d Origin of aorta.
e Arcus aortæ.
f Arteria pulmonalis.
g Left branch (divided).
h Venæ pulmonales sinistræ.
i Ductus arteriosus.
k Aorta descendens.
l Vena cava superior.
m Vena innominata sinistra.
n Arteria iliaca communis.
o Art. iliaca externa.
p Art. iliaca interna.
q Art. umbilicalis.
r Umbilicus.
s Vena umbilicalis.
t Fundus of bladder.
u Urachus.
v Placenta.
w Amnion.
x Chorion.
y Spongy portion of placenta.
z Left lobe of liver.
β Right lobe of liver.
β Gall bladder.
γ Vena umbilicalis.
δ Vena portæ anastomosing with vena umbilicalis.
ε Ductus venosus.
η Vena hepatica.
ϑ Vena cava inferior.
λ Lobulus Spigelii.
μ Kidney.
ν Supra-renal capsule.

Fig. 2.

External Muscles of Larynx.

a Cartilago thyroidea.
b Cartilago cricoidea.
c First ring of trachea.
d Ligamentum crico-thyroideum medium.
e Musculus crico-thyroideus.
f Superior bundle of fibres of musculus crico-thyroideus.

Fig. 3.

Internal Muscles of Larynx.

a Epiglottis.
b Plica ary-epiglottica.
c Cartilago thyroidea.
d Glandulæ arytænoideæ.
e Cartilago Santoriniana.
f Corpusculum Wrisbergii.
g Cartilago cricoidea.
h Ligamentum cemto-cricoideum.
i Musculus crico-arytænoideus posticus.
k Inferior bundle of fibres of musculus crico-arytænoideus posticus attached to inferior cornu of thyroid cartilage.
l Macl. arytænoideus transversus.
m Macl. arytænoideus obliquus.

Fig. 4.

Lateral view of Internal Muscles of Larynx, left wing of Thyroid Cartilage being removed.

a Epiglottis.
b Cartilago thyroidea.
c Cartilago cricoidea.
d Cartilaginous rings of trachea.
e Cartilago arytænoidea.
f Membrana hyo-epiglottica.
g Musculus crico-thyroideus.
h M. crico-arytænoideus posticus.
i M. crico-arytænoideus lateralis.
k Fibres of the preceding which proceed to be attached to the base of the arytænoid cartilage.
l M. thyro-epiglotticus.
m M. thyro-arytænoideus.
n M. arytænoideus transversus.
o M. thyro-arytænoideus superior.
p M. ary-epiglotticus.

Fig. 5.

Posterior view of Interior of cavity of Larynx.

a Posterior surface of epiglottis.
b Os hyoideum.
c Cartilago thyroidea.
d Cartilago cricoidea.
e Upper chordæ vocales.
f Lower chordæ vocales.

g Ventriculus laryngis.
h Musculus thyro-arytænoideus.
i Tracheal opening.

Fig. 6.

Interior of Cæcum or Caput Coli, and termination in it of Small Intestines.

a Ileum.
b Cæcum.
c Colon ascendens.
d Processus vermiformis.
e Opening of vermiform appendix.
f Valvula Bauhini.
g Plicæ sigmoideæ.
h Fossæ of colon.
i Serous coat of small intestine.
k Muscular coat of small intestine.
l Mucous coat of small intestine.

Fig. 7.

Vertical Section of Inferior Lip and Incisor Tooth in Alveolar socket of Lower Jaw.

a Gum attached to neck of tooth.
b Alveolar periosteum.
c Enamel coating.
d Osseous tissue.
e Dental cavity containing pulp of tooth.
f Dental canal.
g Nerve supplying pulp.

Fig. 8.

The Teeth.

a Central incisor, front view.
 α Crown.
 β Neck.
 γ Root or fang.
b Lateral view of root or fang.
c Section of root or fang.
 δ Dental cavity.
 ε Dental canal.
d Canine tooth, front view (cuspidati).
e Canine tooth, lateral view.
f Section of canine tooth.
g Bicuspid tooth, front view (bicuspidati).
h Bicuspid tooth, lateral view.
i Section of bicuspid tooth.
k Molar tooth, front view (molares).
l Molar tooth, lateral view.
m Section of molar tooth.

Tab. XXXVII.

Fig. 3.

Fig. 1.

Fig. 4.

Fig. 2.

Fig. 5.

Fig. 6.

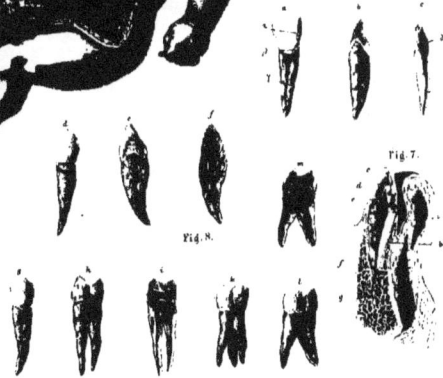

Fig. 8.

Fig. 7.

www.ingramcontent.com/pod-product-compliance
Lightning Source LLC
Chambersburg PA
CBHW021954190326
41519CB00009B/1248